A Guide for Students and Researchers

Statistical Mechanics
made
Simple

A Guide for Students and Researchers

Statistical Mechanics
made
Simple

Daniel C. Mattis

University of Utah, USA

World Scientific

New Jersey • London • Singapore • Hong Kong

Published by

World Scientific Publishing Co. Pte. Ltd.

5 Toh Tuck Link, Singapore 596224

USA office: Suite 202, 1060 Main Street, River Edge, NJ 07661

UK office: 57 Shelton Street, Covent Garden, London WC2H 9HE

Library of Congress Cataloging-in-Publication Data
Mattis, Daniel Charles, 1932–
    Statistical mechanics made simple : a guide for students and researchers / Daniel C. Mattis.
      p. cm.
    Includes bibliographical references and index.
    ISBN 981-238-165-1 -- ISBN 981-238-166-X (pbk.)
    1. Statistical mechanics. I. Title.

QC174.8.M365   2003
530.13--dc21   2003042254

British Library Cataloguing-in-Publication Data
A catalogue record for this book is available from the British Library.

Printed in Singapore by Uto-Print

Contents

Preface

I dedicate this book to those generations of students who suffered through endless revisions of my class notes in statistical mechanics and, through their class participation, homework and projects, helped shape the material.

My own undergraduate experience in thermodynamics and statistical mechanics, a half-century ago at MIT, consisted of a single semester of Sears' *Thermodynamics* (skillfully taught by the man himself.) But it was a subject that seemed as distant from "real" physics as did poetry or French literature. Graduate study at the University of Illinois in Urbana-Champaign was not that different, except that the course in statistical mechanics was taught by the brilliant lecturer Francis Low the year before he departed for... MIT. I asked my classmate J.R. Schrieffer, who presciently had enrolled in that class, whether I should chance it later with a different instructor. He said not to bother — that he could explain all I needed to know about this topic over lunch.

On a paper napkin, Bob wrote "$e^{-\beta H}$". "That's it in a nutshell!" "Surely you must be kidding, Mr Schrieffer," I replied (or words to that effect.) "How could you get the Fermi-Dirac distribution out of THAT? "Easy as pie," was the reply[a]... and I was hooked.

I never did take the course, but in those long gone days it was still possible to earn a Ph.D. without much of a formal education. Schrieffer, of course, with John Bardeen and Leon Cooper, went on to solve the statistical mechanics of superconductors and thereby earn the Nobel prize.

The standard book on statistical physics in the late 1950's was by T. L. Hill. It was recondite but formal and dry. In speaking of a different text that was feebly attempting the same topic, a wit quipped that "it was not worth

---

[a]See Chapter 6.

a bean of Hill's." Today there are dozens of texts on the subject. Why add one more?

In the early 1960's, while researching the theory of magnetism, I determined to understand the two-dimensional Ising model that had been so brilliantly resolved by Lars Onsager, to the total and utter incomprehension of just about everyone else. Ultimately, with the help of Elliot Lieb and Ted Schultz (then my colleagues at IBM's research laboratory,) I managed to do so and we published a reasonably intelligible explanation in *Reviews of Modern Physics*. This longish work — parts of which appar in Chapter 8 — received an honorable mention almost 20 years later, in the 1982 Nobel lecture by Kenneth G. Wilson, who wrote:

"In the summer of 1966 I spent a long time at Aspen. While there I carried out a promise I had made to myself while a graduate student, namely [to work] through Onsager's solution of the two- dimensional Ising model. I read it in translation, studying the field-theoretic form given in Lieb, Mattis and Schultz ['s paper.] When I entered graduate school I had carried out the instructions given to me by my father and had knocked on both Murray Gell-Mann's and [Richard] Feynman's doors and asked them what they were currently doing. Murray wrote down the partition function of the three-dimensional Ising model and said it would be nice if I could solve it.... Feynman's answer was "nothing." Later, Jon Mathews explained some of Feynman's tricks for reproducing the solution for the two-dimensional Ising model. I didn't follow what Jon was saying, but that was when I made my promise.... As I worked through the paper of Mattis, Lieb and Schultz I realized there should be applications of my renormalization group ideas to critical phenomena..."[b]

Recently, G. Emch has reminded me that at the very moment Wilson was studying our version of the two-dimensional Ising model I was attending a large IUPAP meeting in Copenhagen on the foundations and applications of statistical mechanics. My talk had been advertised as, "The exact solution of the Ising model in three dimensions" and, needless to say, it was well attended. I did preface it by admitting there was no exact solution but that — had the airplane taking me to Denmark crashed — the title alone would have earned me a legacy worthy of Fermat. Although it was anticlimactic, the actual talk[c] demonstrated that in 5 spatial dimensions or higher, mean-field theory prevails.

---

[b]From *Nobel Lectures in Physics (1981–1990)*, published by World Scientific.
[c]It appeared in the Proceedings with a more modest title befitting a respectable albeit approximate analysis.

In the present book I have set down numerous other topics and techniques, much received wisdom and a few original ideas to add to the "hill of beans." Whether old or new, all of it *can* be turned to advantage. My greatest satisfaction will be that you read it here first.

D.C.M.
Salt Lake City, 2003

Introduction: Theories of Thermodynamics, Kinetic Theory and Statistical Mechanics

Despite the lack of a reliable atomic theory of matter, the science of Thermodynamics flourished in the 19th Century. Among the famous thinkers it attracted, one notes William Thomson (Lord Kelvin) after whom the temperature scale is named, and James Clerk Maxwell. The latter's many contributions include the "distribution function" and some very useful differential "relations" among thermodynamic quantities (as distinguished from his even more famous "equations" in electrodynamics). The Maxwell relations set the stage for our present view of thermodynamics as a science based on function theory while grounded in experimental observations.

The kinetic theory of gases came to be the next conceptual step. Among pioneers in this discipline one counts several unrecognized geniuses, such as J. J. Waterston who — thanks to Lord Rayleigh — received posthumous honors from the very same Royal Society that had steadfastly refused to publish his works during his lifetime. Ludwig Boltzmann committed suicide on September 5, 1906, depressed — it is said — by the utter rejection of his atomistic theory by such colleagues as Mach and Ostwald. Paul Ehrenfest, another great innovator, died by his own hand in 1933. Among 20th Century scientists in this field, a sizable number have met equally untimely ends. So "*now*", (here we quote from a well-known and popular text[a]) "it is *our* turn to study statistical mechanics".

The postulational science of Statistical Mechanics — originally introduced to justify and extend the conclusions of thermodynamics but nowadays extensively studied and used on its own merits — is entirely a product of the 20th Century. Its founding fathers include Albert Einstein (who, among his many other contributions, made sense out of Planck's Law) and J. W. Gibbs, whose formulations of phase space and entropy basically

---

[a]D. H. Goodstein, *States of Matter*, Dover, New York, 1985.

anticipated quantum mechanics. Many of the pioneers of quantum theory also contributed to statistical mechanics. We recognize this implicitly when-ever whenever we specify particles that satisfy "Fermi–Dirac" or "Bose–Einstein" statistics, or when we solve the "Bloch equation" for the density matrix, or when evaluating a partition function using a "Feynman path integral".

In its most simplistic reduction, *thermodynamics* is the study of mathematical identities involving partial derivatives of well defined functions. These relate various macroscopic properties of matter: pressure, temperature, density, magnetization, etc., to one another. Phase transitions mark the discontinuities of one or more of these functions and serve to separate distinct regions (e.g. vapor from solid) in the variables' phase space. *Kinetic theory* seeks to integrate the equations of motion of a many-body system starting from random initial conditions, thereby to construct the system's thermodynamic properties. Finally, *statistical mechanics* provides an axiomatic foundation for the preceding while allowing a wide choice of convenient calculational schemes.

There is no net flow of matter nor of charged particles in thermody-namic equilibrium. Away from equilibrium but in or near *steady state*, the *Boltzmann equation* (and its quantum generalizations by Kubo and others) seeks to combine kinetic theory with statistical mechanics. This becomes necessary in order to explain and predict transport phenomena in a non-ideal medium, or to understand the evolution to equilibrium when start-ing from some arbitrary initial conditions. It is one of the topics covered in the present text.

Any meaningful approach revolves about taking N, the number of distinct particles under consideration, to the limit $N \to \infty$. This is not such a dim idea in light of the fact that Avogadro's number, $N_A = 6.022045 \times 10^{23}$ *per* mole.[b]

Taking advantage of the simplifications brought about by the law of large numbers and of some 18th Century mathematics one derives the underpin-nings for a science of statistical mechanics and, ultimately, finds a theoretical justification for some of the dogmas of thermodynamics. In the 9 chapters to follow we see that a number of approximate relations at small values of N become exact in the "*thermodynamic limit*" (as the procedure of taking the

---

[b]A *mole* is the amount of a substance that contains as many elementary entities as there are carbon atoms in 12 gm of Carbon 12. E.g.: 1 mole of electrons (e^-) consists of N_A particles of total mass 5.4860×10^{-4} gm and total charge -96.49×10^3 coulombs.

limit $N \to \infty$ is now known in all branches of physics, including many-body physics and quantum field theory).

Additionally we shall study the *fluctuations* $O(\sqrt{N})$ in macroscopic $O(N)$ *extensive* quantities, for "one person's noise is another person's signal". Even when fluctuations are small, what matters most is their relation to other thermodynamic functions. For example, the "noise" in the internal energy, $\langle E^2 \rangle - \langle E \rangle^2$, is related to the same system's heat capacity dE/dT. Additional examples come under the rubric of the "fluctuation-dissipation" theorem.

With Bose–Einstein condensation, "high"-temperature super-conductivity, "*nanophysics*", "quantum dots", and "colossal" magnetore-sistance being the order of the day, there is no lack of contemporary applications for the methods of statistical physics. However, first things first. We start the exposition by laying out and motivating the fundamentals and methodologies that have "worked" in such classic systems as magnetism and the non-ideal gas. Once mastered, these reductions should allow one to pose more contemporary questions. With the aid of newest techniques — some of which are borrowed from quantum theory — one can supply some of the answers and, where the answers are still lacking, the tools with which to obtain them. The transition from "simple" statistical mechanics to the more sophisticated versions is undertaken gradually, starting from Chapter 4 to the concluding chapters of the book. The requisite mathematical tools are supplied as needed within each self-contained chapter.

The book was based on the needs of physics graduate students but it is designed to be accessible to engineers, chemists and mathematicians with minimal backgrounds in physics. Too often physics is taught as an idealized science, devoid of statistical uncertainties. An elementary course in thermo-dynamics and statistical physics can remedy this; Chapters 1–4 are especially suitable for undergraduates aspiring to be theoreticians. Much of the material covered in this book is suitable for self-study but *all* of it can be used as a classroom text in a one-semester course.

Based in part on lecture notes that the author developed during a decade of teaching this material, the present volume seeks to cover many essential physical concepts and theoretical "tricks" as they have evolved over the past two centuries. Some theories are just mentioned while others are developed in great depth, the sole criterion being the author's somewhat arbitrary opinion of the intellectual depth of the posed problem and of the elegance of its resolution. Here, function follows form.

Specifically, Chapters 1 and 2 develop the rudiments of a statisti-cal science, touching upon metastable states, phase transitions, critical exponents and the like. Applications to magnetism and superconductivity are included *ab initio*. Chapter 3 recapitulates thermodynamics in a form that invites comparison with the postulational statistical mechanics of Chapter 4. van der Waals gas is studied and then compared to the exactly solved Tonks' gas. Chapters 5 and 6 deal, respectively, with the quantum statistics of bosons and fermions and their various applications. We distinguish the two principal types of bosons: conserved or not. The notion of "quasiparticles" in fermion systems is stressed. We touch upon semiconductor physics and the rôle of the chemical potential $\mu(T)$ in n-type semiconductors, analyzing the case when ionized donors are incapable of binding more than one excess electron due to 2-body forces. Chapter 7 presents the kinetic theory of dilute gases. Boltzmann's H-function is used to compute the approach to thermodynamic equilibrium and his eponymous equation is transformed into an eigenvalue problem in order to solve for the dispersion and decay of sound waves in gases.

Chapter 8 develops the concept of the transfer matrix, including an Onsager-type solution to the two-dimensional Ising model. Exact formulas are used to calculate the critical exponents of selected second-order phase transitions. The concept of "frustration" is introduced and the transfer matrix of the "fully frustrated" two-dimensional Ising model is diagonalized explicitly. A simplified model of fracture, the "zipper", is introduced and partly solved; in the process of studying this "classical" system, we learn something new about the equations of continuity in quantum mechanics!

Chapter 9, the last, is devoted to more advanced techniques: Doi's field-theoretic approach to diffusion-limited chemical reactions is one and the Green's functions theory of the many-body problem is another. As illustrations we work out the eigenvalue spectrum of several special models — including that of a perfectly random Hamiltonian.

Additional models and calculations have been relegated to the numerous problems scattered throughout the text, where you, the reader, can test your mastery of the material. But despite coverage of a wealth of topics this book remains incomplete, as any text of normal length and scope must be. It should be supplemented by the monographs and review articles on critical phenomena, series expansions, reaction rates, exact methods, granular materials, etc., found on the shelves of even the most modest physics libraries. If used to good advantage, the present book could be a gateway to these storehouses of knowledge and research.

Chapter 1

Elementary Concepts in Statistics and Probability

1.1. The Binomial Distribution

We can obtain all the binomial coefficients from a simple *generating function* G_N:

$$G_N(p_1, p_2) \equiv (p_1 + p_2)^N = \sum_{n_1=0}^{N} \binom{N}{n_1} p_1^{n_1} p_2^{n_2} , \tag{1.1}$$

where the $\binom{N}{n_1}$ symbol[a] stands for the ratio $N!/n_1!n_2!$ of factorials. Both here and subsequently, $n_2 \equiv N - n_1$.

If the p's are positive, each term in the sum is positive. If restricted to $p_1 + p_2 = 1$ they add to $G_N(p_1, 1-p_1) \equiv 1$. Thus, each term in the expansion on the right-hand side of (1.1) can be viewed as a *probability* of sorts.

Generally there are only three requirements for a function to be a probability: it must be non-negative, sum to 1, and it has to express the relative frequency of some stochastic (i.e. *random*) phenomenon in a meaningful way. The binary distribution which ensues from the generating function above can serve to label a coin toss (let 1 be "heads" and 2 "tails"), or to label spins "up" in a magnetic spin system by 1 and spins "down" by 2, or to identify copper atoms by 1 and gold atoms by 2 in a copper-gold alloy, etc. Indeed all non-quantum mechanical binary processes with a statistical component are similar and can be studied in the same way.

It follows (by inspection of Eq. (1.1)) that we can define the probability of n_1 heads and $n_2 = N - n_1$ tails, in N tries, as

$$W_N(n_1) = \binom{N}{n_1} p_1^{n_1} p_2^{n_2} , \tag{1.2}$$

---

[a]Spoken: "N *choose* n_1". Recall $n! \equiv 1 \times 2 \cdots \times n$ and, by extension, $0! \equiv 1$, so that $\binom{N}{N} = \binom{N}{0} = 1$.

subject to $p_2 = 1 - p_1$. This chapter concerns in part the manner in which one chooses p_1 and p_2 in physical processes. These are the parameters that pre-determine the relative *a priori* probabilities of the two events. Of course, by just measuring the relative frequency of the two events one could determine their respective values *a posteriori* after a sufficiently large number of tries N, and on the way measure all other properties of the binary distribution including as its width (second moment), etc.

But this is not required nor is it even desirable. One might attribute $p_1 = p_2 = 1/2$ by symmetry to a perfectly milled coin, *without* performing the experiment. Tossing it N times should either confirm the hypothesis or show up a hidden flaw. Similarly one can predict the width of the binary distribution from theory alone, without performing the experiment.

Thus it becomes quite compelling to understand the consequences of a probability distribution at arbitrary values of the parameters. Experiment can then be used not just to determine the numerical values of the parameters but also to detect systematic deviations from the supposed randomness.

These are just some of the good reasons not to insist on $p_1 + p_2 = 1$ at first. By allowing the generating function to depend on *two* independent parameters p_1 and p_2 it becomes possible to derive all manners of useful (or at least, entertaining), identities. In the first of these one sets $p_1 = p_2 = 1$ in (1.1) and immediately obtains the well-known sum rule for binomial coefficients:

$$\sum_{n_1=0}^{N} \binom{N}{n_1} = 2^N . \tag{1.3a}$$

Setting $p_1 = -p_2$ yields a second, albeit less familiar, sum rule:

$$\sum_{n_1=0}^{N} \binom{N}{n_1} (-1)^{n_1} = 0 . \tag{1.3b}$$

Next, expand $(p + p_2)^{n+k}$ in powers of $p_1^{n_1} p_2^{n_2}$ as in Eq. (1.1) and similarly for each of the two factors $(p_1 + p_2)^n \times (p_1 + p_2)^k$ separately. Upon equating the coefficients term by term one derives the "addition theorem" for the binomial coefficients:

$$\binom{n+k}{r} = \sum_{t=0}^{r} \binom{n}{t} \binom{k}{r-t} . \tag{1.3c}$$

Two special cases of this formula may prove useful. In the first, set $k = 1$, so that t is restricted to $r - 1$ and r. Recalling that $0! \equiv 1$ and $1! = 1$ we deduce $\binom{n+1}{r} = \binom{n}{r} + \binom{n}{r-1}$ from Eq. (1.3c).

In the second example set $k = n = r$. Then (1.3c) yields:

$$\binom{2n}{n} = \sum_{t=0}^{n} \binom{n}{t}\binom{n}{n-t} = \sum_{t=0}^{n} \binom{n}{t}^2.$$

Moreover, by retaining p_1 and p_2 as independent variables in G_N it becomes possible to obtain *all* moments of the distribution simply by differentiating the generating function multiple times, following which $p_1 + p_2 = 1$ is imposed. Let us start by evaluating the lowest moment, i.e. the average of n_1, denoted $\langle n_1 \rangle$ (sometimes also written $\bar{n}_1$).

$$\langle n_1 \rangle \equiv \sum_{n_1=0}^{N} n_1 W_N(n_1) = \left\{ p_1 \frac{\partial}{\partial p_1} G_N(p_1, p_2) \right\}\bigg|_{p_2=1-p_1}$$

$$= \left\{ p_1 \frac{\partial}{\partial p_1} (p_1 + p_2)^N \right\}\bigg|_{p_2=1-p_1} = N p_1 . \tag{1.4a}$$

Similarly,

$$\langle n_1^2 \rangle = \left\{ \left(p_1 \frac{\partial}{\partial p_1} \right)^2 (p_1 + p_2)^N \right\}\bigg|_{p_2=1-p_1} = (N p_1)^2 + N p_1 (1 - p_1) . \tag{1.4b}$$

For higher powers also, $n_1^m \leftrightarrow (p_1 \partial/\partial p_1)^m$ is always the correct substitution. A measure of the width or "second moment" of the distribution may be derived from the *variance* σ, here defined as $\sigma^2 \equiv \langle (n_1 - \langle n_1 \rangle)^2/N \rangle = (\langle n_1^2 \rangle - \langle n_1 \rangle^2)/N$. In the present example, inserting the result (1.4a) in (1.4b) we find $\sigma = \sqrt{p_1 p_2} = \sqrt{p_1(1 - p_1)}$. This result can be put to immediate and practical use.

1.2.   Length of a Winning Streak

In casino gambling, unlike some other real-life situations, persistence definitely does *not* "pay". Take as an example the most favorable situation of a "winning streak". Under the assumption that a coin toss resulting in "heads" wins and "tails" loses, $d = n_1 - n_2$ measures the net winnings (or losses, if negative). With an honest coin, $p_1 = p_2 = 1/2$ and therefore the most probable $\langle d \rangle = 0$. However, by (1.4b), $\langle d^2 \rangle = N/2$. Therefore we estimate the maximum winnings (or losses) after N tries as $\pm$ the "root mean-square" (rms) value (with $d_{\text{rms}} \equiv$ the positive root of $\langle d^2 \rangle$,)

$$d_{\text{rms}} = \sqrt{N/2} . \tag{1.5}$$

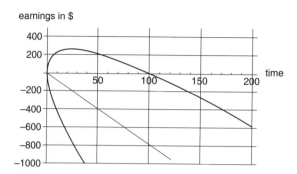

Fig. 1.1.   Examples of most probable gambling streaks.
This plot of "Earnings versus Time spent playing", shows the *most* probable gambling trajectory (according to elementary *biased* RW theory), as a thin straight line. The top curve is the "winning" streak at one standard deviation from the most probable; after an initial winning spurt "in the black" it goes "into the red" after $t > 100$, for the arbitrary parameters used in this illustration. Lower curve (the losing streak), also one standard deviation from the most probable, is negative from the start. Asymptotically *any* reasonably probable trajectory lying between these two curves must end up deeply "in the red".

This assumption yields the most probable *winning streak* although a *losing streak* with $d_{\mathrm{rms}} = -\sqrt{N/2}$ is *equally* probable. In either case the player must contend with the percentage retained by the casino, generally proportional to the number N of plays (or time t spent playing). *However small this percentage might be, ultimately it always exceeds the winnings.* Asymptotically the player loses, as shown in the example of Fig. 1.1 with $N \propto t$. This model is analogous to the one dimensional Brownian motion known as the *biased Random Walk*, to which we return in Problem 1.1.

1.3.   Brownian Motion and the Random Walk

The Scottish botanist Robert Brown observed in 1827 that grains of pollen, coal dust, or other specks of materials in liquid suspension and visible under a microscope, appeared to jump randomly in position and direction. The physical explanation provided by Einstein in 1906 invoked multitudes of invisible molecules striking the visible particles, imparting large numbers of random impulses to them. Let us consider one simplified version of this kinetic theory; a second will follow.

Consider a completely random walker (RW) (either the above mentioned speck or the proverbial "drunken sailor") whose position from the origin after n steps is $\mathbf{r}_n$. Each step is $\mathbf{s}_{n+1} = \mathbf{r}_{n+1} - \mathbf{r}_n$. Assume a given step length s_n and a perfectly random direction $\hat{\mathbf{s}}_n$, each step being uncorrelated with those preceding it. By symmetry $\langle \mathbf{s}_n \rangle = 0$. We define $\lambda^2 = \langle \mathbf{s}_n^2 \rangle$ as the average square step-length. Then if $\mathbf{r}_N$ is the destination,

$$\langle \mathbf{r}_N^2 \rangle = \left\langle \left(\sum_{n=1}^{N} \mathbf{s}_n \right)^2 \right\rangle = N\lambda^2 + 2 \sum_{n=1}^{N-1} \sum_{m=n+1}^{N} \langle \mathbf{s}_n \cdot \mathbf{s}_m \rangle. \qquad (1.6)$$

Due to the lack of correlations all the terms in the double sum can be factored, i.e. $\langle \mathbf{s}_n \cdot \mathbf{s}_m \rangle = \langle \mathbf{s}_n \rangle \cdot \langle \mathbf{s}_m \rangle = 0$, and vanish. Thus the rms distance achieved by RW is $R = \lambda\sqrt{N}$ and lies with equal probability at any point on the surface of a sphere of radius R in three dimensions (3D), on a circle of radius R in 2D and at the two points $\pm R$ in 1D. As the number of steps can be assumed to be proportional to the elapsed time ($N \propto t$) the most probable distance from the origin increases as $R \propto \sqrt{t}$. This power law is recognized as the signature of classical *diffusive* motion, just as $R \propto t$ is the signature of *ballistic* motion.

Problem 1.1. A given biased RW is defined by $\langle \mathbf{s}_n \rangle = \mathbf{a}$ and $\langle s_n^2 \rangle = \lambda^2$, with $\mathbf{a}$ and λ constants. Determine the two rms loci of this biased random walker after N steps in 1D (and cf. Fig. 1.1) and generalize to 2D and 3D, as function of a_x/λ, a_y/λ and in 3D, a_z/λ.

1.4. Poisson versus Normal (Gaussian) Distributions

Both of these well-known statistical distributions can be derived as different limiting cases of the binomial distribution. In this regard the Gamma Function $\Gamma(z)$ and *Stirling's approximation* to $N!$ and $\Gamma(N+1)$ prove useful. First, define

$$\Gamma(z) \equiv \int_0^\infty dt\, t^{z-1} e^{-t} \qquad (1.7)$$

as a function that exists everywhere in the complex z-plane except on the negative real axis. After partial integration on t^{z-1} one finds,

$$\Gamma(z) = z^{-1}\Gamma(z+1), \quad \text{i.e. } \Gamma(z+1) = z\Gamma(z). \qquad (1.8)$$

$\Gamma(1) = 1$ from Eq. (1.7) (by inspection). Hence $\Gamma(2) = 1$, $\Gamma(3) = 2 \cdot 1$ and by induction on any positive integer N, $\Gamma(N) = (N-1)!$ Hence, $\Gamma(1) = 0! = 1$, and $(-N)! = \infty$ by extension. This allows the limits on the sum in Eq. (1.1) to be extended, from $0 \leq n \leq N$, to $-\infty < n < +\infty$, if desired.

Half-integer arguments are of equal importance. To obtain them, first one calculates $\Gamma(\frac{1}{2})$:

$$\left\{\Gamma\left(\frac{1}{2}\right)\right\}^2 \equiv \int_0^\infty dt\, t^{-1/2} e^{-t} \int_0^\infty ds\, s^{-1/2} e^{-s}$$

$$= \int_{-\infty}^\infty dx\, e^{-x^2} \int_{-\infty}^\infty dy\, e^{-y^2}$$

$$= \int_0^{2\pi} d\phi \int_0^\infty dr\, r e^{-r^2} = \pi \tag{1.9}$$

(as obtained by substituting $t = x^2$, $s = y^2$, then switching to radial coordinates). Thus: $\Gamma(\frac{1}{2}) = \sqrt{\pi}$, $\Gamma(\frac{3}{2}) = \frac{1}{2}\sqrt{\pi}$, $\Gamma(\frac{5}{2}) = \frac{3}{2} \cdot \frac{1}{2}\sqrt{\pi}$, etc.

Problem 1.2. Find a *formula* for area of unit sphere $S(d)$ in d dimensions.

$$\left[\text{Hint}: \text{ using } \mathbf{r}^2 = r_1^2 + r_2^2 + \cdots + r_d^2, \text{ compute}: \right.$$

$$\int d^d r\, e^{-\mathbf{r}^2} = S(d) \int_0^\infty dr\, r^{d-1} e^{-r^2} = \frac{S(d)}{2} \int_0^\infty dt\, t^{(\frac{d}{2}-1)} e^{-t}$$

$$\left. = \left\{ \int_{-\infty}^\infty dx\, e^{-x^2} \right\}^d. \right]$$

Evaluate $S(d)$ explicitly for $d = 1, 2, 3, 4, 5$.

Stirling's approximation to $\Gamma(N+1)$ is obtained by setting $t^N e^{-t} \equiv \exp[g(t)]$ and evaluating it by steepest descents. At the maximum of $g(t)$, defined as $t = t_0$, $g'(t) = \partial/\partial t\{-t + N\log t\}|_{t_0} = 0 \Rightarrow t_0 = N$. Approximating $g(t)$ by the first few terms in its Taylor expansion we find $g(t) = g(t_0) + [(t-t_0)^2/2!]g''(t_0) + [(t-t_0)^3/3!]g'''(t_0) + \cdots$, with $g''(t_0) = -1/N$. Third and higher derivatives become negligible in the large N limit. Thus, $g(t) = N\log N - N - (1/2N)(t-N)^2$ is to be inserted into the integral for $\Gamma(N+1)$:

$$\Gamma(N+1) = N! = \exp(N\log N - N) \int_{-N}^\infty dt \exp\left(-\frac{t^2}{2N}\right)$$

$$= \sqrt{2N\pi}\exp(N\log N - N). \tag{1.10}$$

The logarithm of this result yields the more familiar expression, $\log(N!) = N \log N - N + 1/2 \log(2N\pi)$. We note that in most applications the first two terms suffice and the last term is omitted.

Problem 1.3. Estimate the fractional errors in Stirling's approximation arising from the two sources: the neglect of the next term $(1/3! g'''(t - t_0)^3)$ in the expansion of $g(t)$ and the approximation of $-N$ by $-\infty$ in the limits of integration. Compare Stirling's result with the exact values of $\Gamma(z)$ for $z = 3$, 3.5, 10 and 10.5 and obtain the fractional errors numerically. How well do they agree with your estimate?

The *Poisson* distribution is named after the renowned 19th Century mathematician and physicist who, in the Napoleonic wars, was required to analyze the tragic, albeit uncommon, problem of soldiers kicked to death by mules. Was it greater than random? The distribution that bears his name applies to trick coins, radioactive decay of metastable nuclei, and other instances in which some remarkable event being monitored is highly improbable; it is obtained as a limiting case of the binary distribution in the $\lim \cdot p_1 \to 0$. Define $\lambda = Np_1$ as a new, finite, parameter of $O(N^0)$, in the thermodynamic $\lim \cdot N \to \infty$. Thus $n_1 \ll N$ and $n_2 \approx N$. It then becomes permissible to approximate $N!/n_2! \equiv N(N-1)\cdots(N-n_1+1)$ by N^{n_1} and $(1 - p_1)^{n_2}$ by $(1 - \lambda/N)^N \to e^{-\lambda}$ (recall the definition of e). With these substitutions $W_N(n_1) \to P(n_1)$, the Poisson distribution:

$$P(n) = \frac{e^{-\lambda}}{n!}\lambda^n \quad [\text{Poisson}].\qquad(1.11)$$

Remarkably, despite several approximations the normalization is preserved — that is, the sum rule $\sum P(n) = 1$ continues to be satisfied exactly.

The more familiar, i.e. *normal*, distribution is that due to Gauss. The *Gaussian* distribution can also be derived from the binomial whenever the p's are both nonzero and the number $N \to \infty$. Its applications are ubiquitous: the distribution of grades in large classes, of energies in a classical gas, etc.

Because both n_1 and n_2 scale with N the ratios n_1/N and n_2/N can, in some sense, be treated as continuous variables (this is not possible if n_1 is $O(1)$ as in the Poisson distribution). Clearly, $W_N(n_1)$ has its maximum at some $\tilde{n}_1$ which is then defined as the "most probable" value of n_1. As $W_N(n_1)$ must be "flat" at the maximum we look for solution of $W_N(n_1 \pm 1) = W_N(n_1)$.

Using Eq. (1.2), we obtain to leading order in $1/N$:

$$1 \approx \left(\frac{N}{\tilde{n}_1} - 1\right) p_1 (1 - p_1)^{-1}, \quad \text{i.e. } \tilde{n}_1 = p_1 N = \langle n_1 \rangle. \qquad (1.12)$$

In other words, the *most probable* value of n_1 turns out to equal its *average* value. (Similarly for n_2.)

Next, use Stirling's approximation to expand $\log W_N$ about its optimum value. The result (left as an exercise for the reader), is:

$$\log W_N(n_1) = \log W_N(\tilde{n}_1) - \frac{1}{2!} \frac{(n_1 - \tilde{n}_1)^2}{N\sigma^2} + O((n_1 - \tilde{n}_1)^3), \qquad (1.13)$$

using the result obtained in Sec. 1.1, $\sigma^2 = p_1 p_2$. The normal (Gaussian) distribution P follows from the first two terms alone:

$$W_N \rightarrow P(n_1) = \frac{1}{\sqrt{2\pi N\sigma^2}} \exp\left[-\frac{(n_1 - \tilde{n}_1)^2}{2N\sigma^2}\right] \quad \text{[Gauss]}, \qquad (1.14a)$$

or

$$P(x) = \frac{1}{\sqrt{2\pi\sigma^2}} \exp\left[-\frac{(x - \tilde{x})^2}{2\sigma^2}\right], \qquad (1.14b)$$

upon defining $x \equiv n_1/\sqrt{N}$.

Problem 1.4. Derive Eq. (1.13) explicitly and show that the *next* term in the expansion, $O((n_1 - \tilde{n}_1)^3)$, is $C \equiv (p_2^2 - p_1^2)(n_1 - \tilde{n}_1)^3/3! N^2 \sigma^4$. Observe that because (1.14a) effectively restricts n_1 to the range $\tilde{n}_1 \pm O(\sqrt{N})$, $C \rightarrow 0$ in the thermodynamic limit. [*Note:* $C \equiv 0$ if $p_1 = p_2 = 1/2$.] Define $x \equiv (n_1 - \tilde{n}_1)/\sqrt{N}$ as a continuous measure of the fluctuations, in the thermodynamic limit. Then derive $P(x)$ as in (1.14b), especially the lack of a factor $1/\sqrt{N}$, and show it to be normalized, i.e. $\int dx P(x) = 1$. Then calculate $\langle x^2 \rangle$.

Problem 1.5. Using the normal distribution (1.14a) and the result of the preceding Problem, show that:

$$\langle (n_1 - \tilde{n}_1)^k \rangle = \begin{cases} (k-1) \cdots 5 \cdot 3 \cdot 1 \cdot (N\sigma^2)^{k/2} & \text{if } k \text{ is } even, \\ 0 & \text{if } k \text{ is } odd. \end{cases}$$

1.5.   Central Limit Theorem (CLT)

This useful theorem formalizes the preceding results and helps derive the normal distribution without reference to the binomial expansion. Definitions and an example follow.

Generally speaking, the CLT applies to *any* process that results from a large number of small contributions — as in the example of Brownian motion, where the motion of a dust particle is the results of its unobservable collisions with a myriad of smaller, invisible, host molecules. To analyze such situations in more detail we must learn to formulate probabilities for two or more variables.

Let us start with 2 and generalize to N by induction. Let $P(s,t)$ be the probability distribution of 2 independent variables s and t over a finite range. The requirements are that P must be non-negative over that range and its integral normalized, $\int ds \int dt\, P(s,t) = 1$. If one integrates over all t, the remainder is a probability for s. Denote it: $P_1(s) = \int dt\, P(s,t)$. Unless $P(s,t)$ is a symmetric function, P_1 differs from the analogously defined $P_2(t) = \int ds\, P(s,t)$.

Averages over arbitrary functions of the two variables are evaluated in the usual way,

$$\langle f(s,t)\rangle = \int ds \int dt\, f(s,t)P(s,t)\,.$$

Consider $f = g(s)h(t)$ in the special case where s and t are *statistically independent*. Then $\langle f\rangle = \langle g\rangle_1 \langle h\rangle_2$ (the subscripts indicate averages over P_1 or P_2 respectively). This holds for arbitrary g and h iff[b] $P(s,t) \equiv P_1(s)P_2(t)$. The generalization is $P(s,t,u) = P_1(s)P_2(t)P_3(u)$, etc. In the following we examine the x-dependence of a one-dimensional RW in which the end-point $x = s_1 + s_2 + \cdots + s_N$ is the sum over N individual steps, each assumed statistically independent of the others but all governed by the same probability $p(s)$,[c] i.e. $P(s_1,\ldots,s_N) = \prod p(s_j)$.

At this point it is helpful to introduce the *Dirac delta function* $\delta(z)$, a function that is zero everywhere except at the origin where it is infinite — such that its integral is 1. With the aid of this singular function we

---

[b] *If and only if.*
[c] The reader might wish to consider a case in which the p's differ, e.g. suppose the individual $p_n(s_n)$ to depend explicitly on n.

can write,

$$P(x) = \int ds_1 \int ds_2 \cdots \int ds_N \delta\left(x - \sum_{n=1}^{N} s_n\right) P(s_1, s_2, \ldots, s_N)$$

$$= \int ds_1 p(s_1) \int ds_2 p(s_2) \cdots \int ds_N p(s_N) \delta\left(x - \sum_{n=1}^{N} s_n\right). \quad (1.15a)$$

The Dirac delta function has numerous representations, such as the limit of an infinitely high and narrow rectangle of area 1, etc. The one most particularly helpful here is: $\delta(\Delta x) = (1/2\pi)\int_{-\infty}^{+\infty} dk\, e^{ik\Delta x}$. As we saw in Sec. 1.2, the most probable $\Delta x \equiv (x - N\langle s_n \rangle)$ is $O(\sqrt{N})$. The dominant contributions to the integration in (1.15a) are from regions where the product $k\Delta x$ is $O(1)$. Therefore the important values of k are at most $O(1/\sqrt{N}) \to 0$.

$$P(x) = \frac{1}{2\pi} \int_{-\infty}^{+\infty} dk\, e^{ikx}\{Q(k)\}^N$$

where

$$Q(k) = \int ds\, e^{-iks} p(s)$$

$$= 1 - \frac{1}{1!}ik\langle s \rangle + \frac{1}{2!}(-ik)^2\langle s^2 \rangle + \frac{1}{3!}(-ik)^3\langle s^3 \rangle + \cdots. \quad (1.15b)$$

Collecting powers, we evaluate $\log Q$ to leading orders:

$$Q(k) = e^{-ik\langle s \rangle} e^{-(k^2/2)(\langle s^2 \rangle - \langle s \rangle^2)} e^{O(k^3)}. \quad (1.16)$$

The coefficients in the exponent are the so-called *moments* of the distribution (also denoted *cumulants* or *semi-invariants*) at each individual step. We identify the second moment as the variance $\sigma^2 = \langle s^2 \rangle - \langle s \rangle^2$ of an individual step in the above, to obtain:

$$P(x) = \frac{1}{2\pi} \int dk\, e^{ikx} e^{-ikN\langle s \rangle} e^{-(Nk^2\sigma^2/2)} e^{N\cdot O(k^3)}. \quad (1.17a)$$

The integration is rendered more transparent by a change of variables. Let $x \equiv N\langle s \rangle + \xi\sqrt{N}$ where ξ is a measure of the fluctuations of x about its systematic (i.e. *biased*) value $N\langle s \rangle$.

In the present application we find that *all moments beyond the second are irrelevant*. For example, $N \cdot O(k^3)$ in the above exponent is $N \cdot N^{-3/2} \to 0$ throughout the range that contributes most to the integration. Higher moments are smaller yet. Replacing k by the rescaled dummy variable of integration $q = k\sqrt{N}$ we obtain[d] $P(\xi)$:

$$P(\xi) = \frac{1}{2\pi} \int dq\, e^{iq\xi} e^{-q^2\sigma^2/2} e^{N \cdot O(q^3 N^{-3/2})} \xrightarrow[\lim \cdot N \to \infty]{} \frac{1}{\sqrt{2\pi\sigma^2}} e^{-\xi^2/2\sigma^2} . \quad (1.17\text{b})$$

This is the prototype *normal* distribution for the fluctuations. It is independent of $\langle s_n \rangle$. The derivation has preserved the norm: the integral of $P(\xi)$ over ξ in the range $-\infty$ to $+\infty$ remains precisely 1.

The generalization to the 2D or 3D RW *appears* simple, but that may be misleading. With $\mathbf{r} = N\langle \mathbf{s} \rangle + \boldsymbol{\xi}\sqrt{N}$ one readily derives:

$$P(\boldsymbol{\xi}) = \frac{1}{(2\pi\sigma^2)^{d/2}} e^{-\boldsymbol{\xi}^2/2\sigma^2} , \quad \text{in } d \text{ dimensions} . \quad (1.18)$$

However, it remains to define σ^2 appropriately for $d \geq 2$. This is not necessarily straightforward. Consider the two distinct scenarios.

(1) In this model, the Cartesian components at each step, s_x, s_y, ..., are uncorrelated (statistically independent) but subject to $\langle s^2 \rangle = a^2$.

(2) Here the individual step lengths are *constrained* by $s^2 = a^2$, hence the Cartesian components are *not* independent.

The proof of (1.18) and definition of σ in each of these two instances is left as an exercise for the reader. (Hint for #1: the motion in each of the d directions is uncorrelated. For scenario #2: use the three-dimensional Dirac function appropriately.) We conclude this section with some related observations:

- The step distribution needs be neither differentiable nor smooth, for the CLT be valid. For example let each step have equal probability $1/2$ to be $\pm a$, independent of the preceding steps. Then $Q(k) \equiv \cos ka$ and even though $p(s_n)$ is a singular function, Eq. (1.17b) continues to be valid with $\sigma^2 = \langle s^2 \rangle = a^2$.

- Finally, if instead of distance traveled we investigate the momentum $\mathbf{p}$ transferred (or impulses imparted) to an object of mass M by numerous collisions with lighter, invisible molecules of mass $m \ll M$, as in Brown's experiments, we now know (essentially, by inspection) that the result *has to be*

$$P(\mathbf{p}) = \frac{1}{(2\pi\sigma^2)^{3/2}} e^{-\mathbf{p}^2/2\sigma^2} . \quad (1.19)$$

We can turn the variance into a measure of temperature T by defining $\sigma^2 = Mk_{\mathrm{B}}T$, identifying with thermal fluctuations and where k_{B} is the

---

[d]Note: to preserve probability $P(\xi) = P(x(\xi))(dx/d\xi) = P(x(\xi))\sqrt{N}$.

Boltzmann constant. Then the CLT implies the familiar *Maxwell–Boltzmann* distribution in the momentum space of an "ideal" gas of mass M particles. The exponent expresses the energy of the particles in units of k_BT, the thermal unit of energy. Note this derivation does not ensure that any similar expression exists for the momentum distribution of the *unseen* molecules of lighter mass m.[e]

1.6.   Multinomial Distributions, Statistical Thermodynamics

After making some apparently *ad hoc* identifications and generalizations we intend to use the previous results to derive some sort of thermodynamics from "first principles". This section is designed to motivate the following, more rigorous, chapters.

Assume particles carry r distinguishable labels. For example, $r = 2$ for the coin toss, 6 for dice and $2S+1$ for spins S ($S = 1/2, 1, 3/2, \ldots$) The total number of particles is then $N = \sum n_j$, where n_j is the number of particles with label j running from 1 to r. Independent of any *a priori* probabilities p_j the statistical factors are "N choose $n_1, n_2, \ldots$". The multinomial probability distribution is,

$$V_N(n_1, n_2, \ldots, n_r) = N! \prod_{j=1}^{r} \frac{p_j^{n_j}}{n_j!} . \tag{1.20}$$

These are all positive quantities that can be obtained from the expansion of a generalized *generating function* $G = (\sum p_j)^N$. Therefore if the p_j add up to 1, $G = 1$ and the V's are also normalized.

The V's are sharply peaked about a maximum. Our experience with the binomial distribution has shown that even though V is highly singular its *natural logarithm* can be expanded in a Taylor series about the maximum. Let us first examine this maximum.

$$\log V_N = \left[\sum_{j=1}^{r} n_j \log(p_j) \right] + \left[\log(N!) - \sum_{j=1}^{r} \log(n_j!) \right] . \tag{1.21a}$$

We then *arbitrarily* identify the first bracket on the *rhs* of the equation (which involves the parameters p_i), with the negative of the energy of the system. Similarly we identify the second, purely statistical, bracket with the

---

[e] Just as the Gaussian $P(\xi)$ does not mirror arbitrary $p(s)$ in Eqs. (1.15)–(1.17).

product of its entropy and the temperature.[f] Then, the above takes the form,

$$[\log V_N] = \beta[-E] + \beta[T\mathscr{S}]\,. \tag{1.21b}$$

The overall-factor β has units $[\text{energy}]^{-1}$, introduced to make the expression dimensionless. Each bracketed quantity in (1.21a) is *extensive* ($\propto N$), hence E and $\mathscr{S}$ are also extensive. This requires the temperature T to be *intensive*, i.e. independent of N.

The signs have been chosen such that the maximum V behaves properly in two known limiting cases. It must correspond to an *energy minimum* (i.e. the "virtual forces" all vanish) when $\mathscr{S}$ and T are held constant, and with *maximal entropy* at constant E and T.

Similarly let us identify the LHS of the equation with $-\beta F$, where F is defined as a "free energy". Upon dividing by β we recapture the well-known thermodynamic relation: $F = E - T\mathscr{S}$. (Still lacking is identification of β as $1/k_BT$.) Thus, maximizing V is the same as *minimizing the free energy* F.

To minimize the free energy we make use of Stirling's approximation in (1.21a) and optimize *w.r.* to each n_j (treated as a continuous variable.)

$$\frac{\partial}{\partial n_j}\left\{n_j\log p_j - n_j\left[\log\left(\frac{n_j}{N}\right) - 1\right]\right\} = 0\,, \quad \text{for } j = 1,\ldots,r\,. \tag{1.22}$$

Note that we can add any arbitrary multiple of $(N - \sum n_j) = 0$ to the expression in curly brackets.

The solution of (1.22) yields the most probable values of the n_j's, denoted $\tilde{n}_j$ as before. We observe once again that the most probable value is also the average value: $\tilde{n}_j = p_jN = \langle n_j \rangle$.

In the mantra of kinetic theory (and of statistical mechanics) the so-called *ergodic hypothesis* occupies a place of honor. Crudely put, it states that any system permitted to evolve will ultimately, or asymptotically, tend to a state of maximal probability — beyond which it must cease to evolve. One makes the connection with thermodynamics by identifying the *most probable configurations* with those found in *thermodynamic equilibrium*. In the axiomatic statistical mechanics outlined in Chapter 4, this common-sense notion is elevated to a high principle.

Here, to make a more explicit connection with elementary thermodynamics, we identify the *a priori* probabilities with Boltzmann factors, $p_j = \exp[-\beta(\varepsilon_j - \mu)]$. The quantity ε_j is the energy of a particle in the jth state and the "chemical potential" μ is chosen to allow p_j to satisfy $\sum p_j = 1$.

---

[f]These quantities will be given a rigorous definition in the next chapter.

Some recognizable results can be found upon using this p_j in Eq. (1.21a). For example, the statistically averaged (also, the most probable) internal energy is

$$N \sum_{j=1}^{r} e^{-\beta(\varepsilon_j - \mu)}(\varepsilon_j - \mu) \equiv E - \mu N ,$$

in which E is defined as the total averaged "physical" energy. We shall find similar expressions in connection with the theory of ideal gases.

1.7.   The Barometer Equation

To show that the identification of the p_j with the energy exponential was not entirely capricious we now derive the well-known law for the drop in barometric pressure with altitude.

Consider an (imaginary) vertical tube of cross-section A in the atmosphere, across which we stretch two horizontal membranes to measure the pressure at altitudes z and $z + dz$. Clearly, the drop in pressure is:

$$dp = -mg\rho dz , \tag{1.23}$$

where ρ is the particle density at z. The ideal gas law $p = \rho k_B T$ is useful in this context. Replacing ρ by $p/k_B T$ in this relation we obtain:

$$p = p_0 e^{-mgz/k_B T} \tag{1.24}$$

where p_0 is the pressure at the reference surface, $z = 0$.

This suggests that both the kinetic *and* the potential energies are distributed exponentially in the dilute or ideal gas. This remarkable result is also quite reasonable, considering that either form of energy can always be transformed into the other, but a *proof* requires statistical mechanics as developed in a later chapter.

1.8.   Other Distributions

There are instances in which the *lognormal* distribution plays a role, in which it is the logarithm of the random variable that is normally distributed. Among other, more exotic distributions, we count the *stretched exponential*, the *Lorentzian*, and numerous others for which the moments do or do not exist, as covered in numerous texts on mathematical statistics.[g] We deal with some of them elsewhere in this book. But with the sole exception of the "Lorentzian", $P(x) = \frac{\gamma/\pi}{x^2+\gamma^2}$, what makes them exotic is their rarity in physical applications, in comparison with either the Gauss or Poisson distributions. On the other hand, the far more esoteric *stochastic matrices* have found their niche in physics ever since their introduction by Wigner,[h] who approximated the Hamiltonian of the quantum nuclear many-body problem by a random matrix.

It is helpful to visualize a random matrix as some kind of a quantum-mechanical RW in which the nth step connects not just the nth position of the random walker to the $n+1$st, but also to all other positions that he ever has or ever will occupy. One may denote this process a "multiply-connected" RW and represent it by an $N \times N$ real, symmetric, matrix:

$$
\overset{\approx}{M} = \begin{pmatrix} m_{11} & m_{12} & \cdots & \cdots \\ m_{12} & m_{22} & \cdots & \cdots \\ \vdots & \vdots & \ddots & \\ \vdots & \vdots & & m_{NN} \end{pmatrix}.
\tag{1.25}
$$

In one well-known example, the individual matrix elements m_{ij} are real, random and normalized as follows: $m_{ij} = \varepsilon_{ij}/\sqrt{N}$, with $\langle \varepsilon_{ij} \rangle = 0$ and $\langle \varepsilon_{ij} \rangle^2 = \sigma^2$. All $N(N+1)/2$ matrix elements on or above the main diagonal are statistically independent. One can prove that such a matrix has N real eigenvalues λ_n which, in the *thermodynamic limit* $N \to \infty$, are smoothly distributed according to a distribution function $P(\lambda)$ in the interval $-2\sigma < \lambda < +2\sigma$. As in the CLT, the global result does not depend on the distribution of the $p(\varepsilon)$ of the individual random variables. So one can assume for the latter whichever is more convenient: the binomial distribution (each independent $\varepsilon_{ij} = \pm\sigma$ with equal probability) or the continuous Gaussian ensemble, $P(\varepsilon_{ij})$ with $\sigma = 1$.

[g]See, e.g., M. Evans, N. Hastings and B. Peacock, *Statistical Distributions* (Wiley, New York, 1993). This 2nd edition describes 39 "major" distributions.
[h]See M. L. Mehta, *Random Matrices* (Academic, New York, 1967).

Problem 1.6. (a) Using a random number generator for the individual ε's, use your PC to numerically diagonalize the matrix M in Eq. (1.25), specifying $N = 101$. List the eigenvalues λ_n in ascending order. Plot the level spacings $\lambda_{n+1} - \lambda_n$ as a function of λ_n.

(b) Estimate how large N must be in order that $P_N(\lambda)$ approach its asymptotic limit function $P(\lambda)$ to within $\pm 1\%$ everywhere in the interval $-2 < \lambda < +2$. (Note: the analytical form of $P(\lambda)$ is derived in a later chapter of this book.)

Chapter 2

The Ising Model and the Lattice Gas

2.1. Some Background and Motivation

The formulation of the statistical theory of fluids poses many formidable challenges, including that of disentangling the geometry of 3- and 4-body collisions. The theory of magnetism poses a different but no less daunting set of difficulties, mainly those associated with the commutation relations of spin operators. Amazingly, there does exist one highly simplified approach that can be used to study both magnetism and fluid dynamics. In the context of magnetism it is known as the *Ising model*. The elementary spins are discrete vector-like entities localized on lattice sites. They point "up" or "down" or, equivalently, are quantified as ± 1. In the study of defective solids and fluids the same theory goes by the name of *lattice gas model*: atoms are present $(+1)$ or absent (-1) on given lattice sites. In both applications, any lack in realism is amply compensated by an utter simplicity. In this chapter we shall develop an informal but plausible version of statistical thermodynamics on the basis of this model.

It is not important that this theory accurately fit existing magnetic or crystalline materials, although it does have numerous such physical applications. Here it is used as a pedagogical device to introduce key concepts of thermodynamics and statistical mechanics. Historically, the Ising model has had its greatest success not in physics but in such far-flung fields as sociology, genetics, economics, etc. Moreover, with only slight generalizations (e.g. "Potts model") it has been used to model multi-component alloys and establish their stability diagrams, to simulate the process of crystal growth from the melt, to quantify the surface roughness of materials, to analyze the spread of forest fires, etc., etc.

17

2.2. First-Principles Statistical Theory of Paramagnetism

Let us consider N atomic spins $s_j = \pm 1$, each carrying an elementary magnetic dipole moment $\mathbf{m}_0$ immersed in a common external magnetic field B. The energy of a spin parallel (+ or "up") to the external field is $-m_0 B$, that of an antiparallel spin is $+m_0 B$. The unit of energy is thus $e_0 \equiv m_0 B$. Assume n_1 spins are up and $n_2 = N - n_1$ are down. If one neglects the electromagnetic- and exchange- interactions among spins (assuming they are too distant from one another to couple effectively), the total energy comes from interactions of each dipole with the external field:

$$E(x) = -N(2x - 1)e_0, \quad \text{where } x = n_1/N. \tag{2.1}$$

The energy is a function of x in the interval $0 < x < 1$. In the thermodynamic limit $N \to \infty$ the *average* energy of a spin $\varepsilon(x) = E/N$ becomes a continuous function of x even though any *given* spin has only the two possibilities $\pm e_0$. The total magnetization $M(x) = N\mathbf{m}_0(2x-1)$ also appears to be proportional to the energy in (2.1), but this proportionality is pure coïncidence and disappears once mutual interactions amongst the dipoles is taken into account.

We call E and M "*extensive*", to distinguish them from *intensive* quantities such as B or x, which do not depend on N, the size of the system. One can always determine which variables are extensive and which are intensive by observing how they scale when the size of the system is doubled.

Most often an intensive *force* and an extensive *quantity* exist as conjugate pairs. B and M are such a pair, as are μ (the "chemical potential") and N, or p and V (pressure and volume). We shall examine properties of such conjugate variables in due course.

In order for a thermodynamic equation to make sense all terms have to carry the same units. What is more, they must all scale with the same power of $N-$ or else in the thermodynamic limit $N \to \infty$ the equation becomes ill defined.

To proceed with the Ising paradigm we seek a physical principle with which to specify the *a priori* probabilities p_1 and $p_2(= 1 - p_1)$ of a given spin being up or down. There were some hints at the end of the preceding chapter. In non-relativistic dynamics, whether classical or quantum, the *energy* is a privileged constant of the motion. In dynamical equilibrium it is a minimum. *Thermo*dynamically, regardless of energy, one can only access those configurations that are most probable. The reconciliation of these two requirements provides a framework in which to construct the model statistical thermodynamics of this chapter.

By analogy with W_N let us start by constructing the probability $Q_N(n_1)$ of n_1 spins being up. It is the product of two factors: P that depends solely on the energy and W that depends solely on statistical probability, $P \times W$. This (un-normalized) probability to accompany Eq. (2.1) is

$$Q_N(x) = P(E) \times \binom{N}{n_1}, \quad \text{where } x = \frac{n_1}{N}. \tag{2.2}$$

The statistical factor strongly favors the most numerous, i.e. most probable, configurations at $x = 1/2$ in the present instance. On the other hand, P should be so constructed as to favor the lowest or ground state energy. In the present model, $-E$ peaks at $x = 1$, just where the statistical factor is smallest! Clearly, the *joint* probability $Q_N(x)$ peaks at an intermediate value. To determine the optimal x approximate $\log Q$ with the aid of Stirling's formula. From (1.21a),

$$\log Q_N(x) = \log P(E(x)) - N[x \log x + (1-x) \log(1-x)]. \tag{2.3}$$

Because the last term is extensive, $\log Q$ and $\log P$ must both be extensive as well. P has to be defined so as to encourage low energies and discourage high energies. The most logical candidate for $\log P$ is therefore $-\beta E(x)$, i.e. $-N\beta\varepsilon(x)$, in which we introduce a parameter β which is intensive and carries units of energy^{-1} designed to render $\log P$ dimensionless.

Similarly, we re-label the left-hand side of this equation $\log Q_N(x) \equiv -\beta F(x) \equiv -N\beta f(x)$. If Q_N is to be a *maximum* in *thermodynamic equilibrium*, F and f must be at a *minimum*.[a] We give the "free energy functions" $F(x)$ and $f(x)$ special names once they are optimized with respect to all variables.[b] At, and *only at*, its minimum, is F denoted the "*free energy*". The corresponding quantity $f = F/N$ is then the free energy *per* spin.

Minimizing (2.3) *w.r.* to x yields the most probable value of x,

$$\tilde{x} = \frac{1}{1 + e^{-2\beta e_0}}, \tag{2.4}$$

---

[a]The proof is by contradiction: if Q were not *initially* a maximum the system would evolve in time until ultimately it reached maximum probability- i.e. thermodynamic equilibrium.
[b]Note: in the present model there is only a single variable, x.

labeled by a tilde, as before. Insert this in Eq. (2.3) to obtain an explicit expression for the free energy:

$$F = F(\tilde{x}) = -\frac{N}{\beta} \log[2 \cosh \beta e_0] \,. \tag{2.5}$$

An expansion, e.g. $Q_N(x) = \exp\{-\beta F(\tilde{x}) - A(x - \tilde{x})^2 + \cdots\}$, used as an (un-normalized) probability function, makes it possible to calculate fluctuations of thermodynamic quantities about their optimal values. (We address this interesting side issue in Problem 2.2 below.) Finally, note that if we replace B by $-B$, $x \to 1 - x$. However, the free energy in (2.5) remains invariant and is therefore a "scalar" (as are such other quantities such as T and β under this transformation).

Problem 2.1.   Show that the equilibrium magnetization in this model is $\langle M \rangle = N m_0 \tanh \beta e_0$.

Problem 2.2.   (A) Show that the variance in the number of spins up is $\langle (n_1 - \tilde{n}_1)^2 \rangle / N = \frac{1}{4 \cosh^2 \beta e_0}$ and thus: $\langle (M(n_1) - \langle M \rangle)^2 \rangle$ is extensive. (B) Is $\langle (F(n_1) - F)^2 \rangle$ also extensive? Calculate this using $Q_N(x)$ and discuss.

We have not yet named the statistical contribution in Eq. (2.3), $-N[x \log x + (1 - x) \log(1 - x)]$. It is a dimensionless entity that we now identify with *entropy* and henceforth denote $k^{-1}\mathscr{S}$, where k is Boltzmann's constant. After replacing it in Eq. (2.3), the equation now reads: $F = E - (\beta k)^{-1}\mathscr{S}$. Identifying $(\beta k)^{-1}$ *with temperature T* yields the thermodynamic expression, $\delta F = \delta E - T \delta \mathscr{S}$.

Still, T is *not* arbitrary but is determined by the condition that F be minimized. Setting $\delta F = 0 \Rightarrow T = \delta E / \delta \mathscr{S}|_{\text{opt}}$. We may take this to be the fundamental definition of temperature. As a result, T is intensive and generally restricted to positive values because, under most (albeit not all!)[c] circumstances, the entropy increases monotonically with increasing energy.

---

[c]The remarkable concept of *negative* temperature applies to weakly interacting spins. If a strong external magnetic field is rapidly reversed the lowest energies become the highest and the derivative defining T changes sign. But this is only *quasi*-equilibrium. Once the spins have time to rotate to the new orientation, true thermodynamic equilibrium is restored and T becomes positive once again.

2.3. More on Entropy and Energy

Given $N(E)$ as the number of configurations of energy E, the entropy is quite generally defined as $\mathscr{S} = k \log N$. It is an extensive quantity and is naturally conjugate to the temperature T. But the phase space at E has to be discrete in order to be countable. In the late 19th Century J. Willard Gibbs had already found it necessary to discretize the phase space in classical statistical mechanics so as to limit the magnitude of $\mathscr{S}$ in the ideal gas. Although done for purely technical reasons, this discretization uncannily anticipated quantum mechanics.

But even where counting is possible it remains a tedious task. It may be impractical to measure (or sometimes even to calculate) $\mathscr{S}$ directly in non-trivial models of interacting particles or spins, but one can always deduce it from $\mathscr{S} = (E - F)/T$ or from $\mathscr{S} \equiv -\partial F/\partial T$.

Finally, because F is stationary the partial derivative equals the total derivative and $\mathscr{S}$ is *also* given by $\mathscr{S} = -dF/dT$. Similarly, $E \equiv \partial(\beta F)/\partial\beta = d(\beta F)/d\beta$. Thus the energy E and the entropy $\mathscr{S}$ are both *first derivatives* of a stationary F. We exploit this property when deriving helpful thermodynamic identities in the next chapter.

2.4. Some Other Relevant Thermodynamic Functions

The most salient feature of the model under scrutiny is its magnetization, also an "order parameter" of sorts. In Problem 2.1 it was found that when the optimal x is inserted into the expression for $\langle M \rangle$ the result was $\langle M \rangle = Nm_0 \tanh \beta e_0$. The same result is obtained by differentiating F in (2.5) using the following definition:

$$M(B,T,\ldots) \equiv \frac{-\partial F(B,T,\ldots)}{\partial B}, \qquad (2.6a)$$

which can be integrated to yield

$$F(B,T) = F(0,T) - \int_0^B dB' M(B',T), \qquad (2.6b)$$

in which $F(0,T) = F_0(T)$ is the free energy in zero external field. In the present model, as well as in many non-ferromagnetic real materials, $M(B)$ vanishes at $B = 0$ and is proportional to B at small fields. This linear dependence is written $M(B) = \chi_0 B$, with the constant of proportionality or *zero-field magnetic susceptibility* χ_0 being positive in *paramagnetic* substances (these are the most common) and negative in *diamagnetic* materials

(including some molecules and semi-metals but, more typically, superconductors). In weak fields, then, the free energy is

$$F \approx F_0 - \frac{1}{2}\chi_0 B^2 . \tag{2.6c}$$

The preceding expression makes it clear that paramagnetic substances are
attracted to the source of magnetic fields and diamagnetic ones are repelled,
as should be apparent from their Greek etymology. More generally χ itself
is a function of B, T, and other parameters. In arbitrary field B it is defined
by $\chi(B, T, \ldots) \equiv \frac{\partial M(B,T,\ldots)}{\partial B}$, revealing it as a *second* derivative of F.

The *heat capacity* is another such thermodynamic function. Again,
because F is stationary, C can be written in either of two ways:

$$C = \frac{dE}{dT} = T\frac{d\mathscr{S}}{dT} . \tag{2.7}$$

But $\mathscr{S} \equiv -dF/dT$, hence $C = -T(d^2F/dT^2)$. Therefore the heat capacity
is also a second derivative of F. Later we prove under the most general
conditions that C and $\mathscr{S}$ are always positive, therefore F is a monotonically
decreasing convex function of T.

Both the susceptibility and the heat capacity are extensive quantities. It
is convenient to divide each by N to obtain *intrinsic* quantities, independent
of the amount of material. The ratio $C \div N$ (or $C \div Volume$, depending on
context) is just such an intrinsic property. Denoted *specific heat*, it is written
with a lower-case c. The *specific* susceptibility $\chi \div N$ is another intrinsic
property, frequently written also as χ without much risk of confusion.

In future applications it will be important to note that *all* physical properties intrinsic to a material or to a mathematical model of a material
(its specific heat, electrical and magnetic susceptibilities, dielectric constant,
compressibility,..., etc.) can be expressed as second derivatives of the free
energy. For free spins in an external field the two most relevant such quantities are:

$$c(B,T) = k\frac{m_0 B/kT)^2}{\cosh^2(m_0 B/kT)} \quad \text{and} \quad \chi(B,T) = \frac{m_0^2/kT}{\cosh^2(m_0 B/kT)} . \tag{2.8}$$

In the $\lim \cdot B \to 0$ at finite T, the heat capacity c vanishes identically while
$\chi \to \chi_0 = m_0^2/kT$. (This is *Curie's Law*). On the other hand at any finite
value of B, however small, c is always positive and peaks at $T \propto m_0 B/k$.
In the $\lim \cdot T \to 0$, $\chi(B,0) = 2m_0\delta(B)$ (using the delta function previously
defined in connection with Eq. (1.15)). The *singularity* at $B = T = 0$ identifies the *critical point* of the free-spin model; internal interactions may push

the singularity to finite T, transforming it into a true order-disorder *phase transition*. We examine this phenomenon next.

2.5. Mean-Field Theory, Stable and Metastable Solutions

One century ago P. Weiss introduced the insightful (but approximate) concept of a *molecular field*, a concept so useful that it has been adapted to many-body quantum physics, quantum field theory and other applications too numerous to mention. His approach, now generally known as *mean-field theory*, was itself based on a reading of van der Waals' derivation of an equation of state for a non-ideal gas (cf. Chapter 3), although of two parameters used by van der Waals, Weiss needed only one.

He supposed each spin to be subject not only to the external field B_{ext} but, additionally, to an internally generated field B_{int} simulating the physical interactions with all the other spins. This internal field is related to an *order parameter* which, for spins, is defined as $\sigma \equiv M/Nm_0$ (formerly denoted $2x - 1$). Let $B_{\text{int}} \equiv B_0\sigma$, where B_0 is a parameter measuring the strength of the internal forces — i.e. it is a physical property of the system. The expression for the magnetization in Problem 2.1 is now generalized to,

$$M = Nm_0 \tanh \frac{m_0(B_{\text{ext}} + B_{\text{int}})}{kT} \quad \text{or}, \quad \sigma = \tanh \frac{m_0(B_{\text{ext}} + \sigma B_0)}{kT}. \quad (2.9a)$$

Expressing B_{ext} in units of B_0 (achieved by setting $B_{\text{ext}} \equiv bB_0$) results in the following dimensionless equation, plotted overleaf:

$$\sigma = \tanh \frac{T_c}{T}(b + \sigma). \quad (2.9b)$$

When rotated by $90°$ the above figure is a plot of $\sigma(b, t')$ as a function of $t' = T/T_c$ at various values of b. It has both positive and negative solutions. For the physically stable solutions the sign of σ is that of b; there are no other adjustable parameters.

Metastable solutions (those in which $\sigma b < 0$) can play a role in the context of "spin resonance". In NMR (nuclear magnetic resonance) or ESR (electron spin resonance) it is common to invert an external magnetic field (rotate it by $180°$) in a time rapid compared with the inherent relaxation time of the magnetic system. The spins are then out of equilibrium and their precession towards a new equilibrium yields useful information on the internal forces. The change of external $b \to -b$ at constant σ is equivalent, for weakly interacting spins, to imposition of a *negative* temperature. We can also infer this from Fig. 2.1.

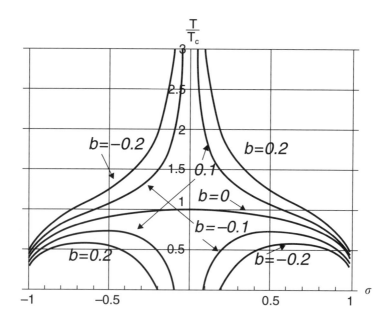

Fig. 2.1.   Mean-field ferromagnetic phase diagram.

Plot of $t' \equiv T/T_c$ (*vertical axis*) versus σ, the order parameter or magnetization *per* site (*horizontal axis*), according to the mean-field theory, Eq. (2.9). The curve labeled $b = 0$ is the *only one* which is continuous over the entire range $-1 < \sigma < +1$; it marks a true thermodynamic phase transition ("critical point") at the point $\sigma = 0$, $t' \equiv T/T_c = 1$.

In nonzero external field the 4 curves $b = \pm 0.1$, ± 0.2 above the one labeled $b = 0$ fail to intersect the T axis, showing there can be no phase transition for $b \neq 0$ at any finite T. The 4 curves *below* $b = 0$ are unstable (i.e. metastable) solutions of Eq. (2.9) that correspond to a *maximum* in the free energy.

Exercise for the reader: extend this diagram to negative temperatures $t' < 0$, either by extrapolation of the above figure or by numerical calculation.

But how is σ to be kept constant? From Eq. (2.9b), the present "*equation of state*", we extract the dependence of external field on $t' = T/T_c$ at σ constant:

$$b(t') = \sigma \left(\frac{t'}{\tau(\sigma)} - 1 \right), \quad \text{with } \tau(\sigma) \equiv \frac{2\sigma}{\log\left(\frac{1+\sigma}{1-\sigma}\right)}$$

(Note: $\lim_{\sigma \to 0} \cdot \tau \longrightarrow 1$.) Thus in an adiabatic process b has to be lowered linearly with t', with slope $\sigma/\tau(\sigma)$. (Lowering it too far, into the range $t' < \tau$ causes b to have a sign opposite to σ and brings the system into a *metastable* regime — as shown in the figure and discussed in the caption.)

Thus the way to induce long-range order in a ferromagnet is to start by heating it above T_c in a magnetic field. Not a new idea! A prescription for the manufacture of compass needles, in Chapter XII of one of the earliest texts in modern physics, William Gilbert's *De Magnete, ca.* 1600, states: "...let the blacksmith beat out upon his anvil a glowing mass of iron...[while] standing with his face to the north...better and more perfectly in winter in colder air when the metal returns more certainly to its normal state..."

In zero external field the $\pm$ branches meet at $T = T_c$. Because the symmetry of either branch is lower than that of the model, either of these solutions demonstrates *spontaneous* symmetry breaking, a phenomenon associated with the spontaneous generation of *long-range order* for $T < T_c$ (in which the majority of spins are parallel, being "up" or "down").

The "critical" properties of the model near T_c are obtained by expanding in $t \equiv (T - T_c)/T_c = t' - 1$; σ and b are *also* assumed to be small or zero. Expanding Eq. (2.9b) near T_c one obtains in leading order,

$$\sigma - \frac{1}{1+t}\left[\sigma + b - \frac{1}{3}(\sigma + b)^3 + \cdots\right] = 0.$$

This has roots:

$$\left.\begin{array}{c}
\sigma = \pm(-3t)^{1/2} \text{ (for } b = 0 \text{ and } t < 0), \\[2mm]
\sigma = \mathrm{sgn}(b) \times (3|b|)^{1/3} \text{ at } t \equiv 0, \quad \text{and} \\[2mm]
\sigma/b \xrightarrow[b\to 0]{} \chi_0 = t^{-1} \text{ for } t > 0.
\end{array}\right\} \qquad (2.10)$$

Near the critical point, the leading terms are in agreement with the solutions plotted in Fig. 2.1. Two *critical indices* denoted β and δ are conventionally used to fit σ to power laws: $\sigma \propto (-t)^\beta$ (for $t < 0$ and $b = 0$) and $\sigma \propto \mathrm{sgn}(b)|b|^{1/\delta}$ (at $t \equiv 0$.) In Weiss' model, according to Eq. (2.10) the critical exponents are $\beta = 1/2$ and $\delta = 3$ respectively, their so-called "classical values". A third critical exponent α measures the singularity in the specific heat at T_c at $b = 0$, through $c \propto 1/|t|^\alpha$. When $b = 0$ in the present model, $E \propto -\sigma^2 \propto t$ for $t < 0$ and $E = 0$ for $t > 0$. Hence c is discontinuous at $t = 0$, with $\alpha = 0$ being the (best[d]) fit to the classical specific-heat exponent. A fourth critical exponent γ serves to fit the zero-field susceptibility χ_0 to a power-law, $\chi_0 \propto t^{-\gamma}$. From Eq. (2.10) we deduce $\gamma = 1$ in the mean-field theory.

---

[d]$\alpha = 0$ is the "best" fit to a power-law in two cases: piece-wise discontinuous functions or functions that are logarithmically divergent.

The critical exponents $\alpha = 0$, $\beta = 1/2$, $\gamma = 1$ and $\delta = 3$ just derived are typical *mean-field* exponents and agree with detailed calculations on more realistic models in $d \geq 4$ dimensions. In $d = 2$ or 3 dimensions the mean-field predictions are typically less reliable, and in $d = 1$ they are flat-out wrong. (In Secs. 2.7 and 2.8 we work out one-dimensional examples in which mean-field theory fails explicitly). Nonetheless the mean-field may provide a useful starting point for more sophisticated theories in which fluctuations are taken into account explicitly; under some circumstances it may be possible to "patch in" the fluctuations so as to correct the mean-field results in $d \leq 4$ dimensions.

2.6.   The Lattice Gas

Imagine a three-dimensional grid dividing space into N cubes of volume a^3 each capable of accommodating 1 atom at most. The volume is Na^3, the number of atoms $N_{at} \leq N$, and the particle density $\rho = N_{at}/N$. As we have seen in Chapter 1, in the absence of forces the statistical factor $\binom{N}{N_{at}}$ stands alone and peaks at $\rho = 1/2$, half occupancy.

Let us assign spin "up" to an occupied site $s_n = +1$ and spin "down" $s_m = -1$ to an unoccupied (vacancy) site. That identifies the local particle density as $\rho_n = \frac{1+s_n}{2}$ and its average as $\rho = \frac{1+\langle s_n \rangle}{2}$.

A one-body energy term $H_1 \equiv -\mu \sum \rho_n$ can serve to adjust the value of ρ from 0 to 1. At $\mu = 0$ the most probable density is $1/2$. Thus $2\rho - 1$ is the quantity analogous to σ while μ, the *chemical potential*, is analogous to $B_{ext}/2$. For stability we need a two-body *interaction* energy. Its mean-field value is $H_2 \equiv -\frac{U}{2N}N_{at}^2 = -\frac{U}{2}N\rho^2$, which is tantamount to modifying μ by an internal field $\mu_{int} = -U\rho$. This completes the analogy of the mean-field lattice gas to the Curie–Weiss model of magnetism. We can use it to derive an equation of state.

2.7.   The Nearest-Neighbor Chain: Thermodynamics in 1D

In one dimension the mean-field approach fails, but there are other approaches that work well. We examine a one-dimensional ferromagnet originally studied by Ising. First, set $b = 0$.

Here the energy of the nth bond is $H_n = -Js_ns_{n+1}$. As before, each s_n can assume the two values ± 1 and n ranges from 1 to N. In the absence of an external field the total energy is $H = \sum H_n$. For present purposes we can calculate the thermodynamic properties by transforming the initial set of coördinates as follows: $s_1 \to \tau_1$, $s_2 \to \tau_1\tau_2$, $s_3 \to \tau_1\tau_2\tau_3$, etc. Because $(\tau_n)^2 = 1$, $H_n \to -J\tau_{n+1}$. Thus the model reduces to $N-1$ "τ" spins in a uniform external field J. Using the previous results, Eq. (2.5), we obtain without further calculations,

$$F = -(N-1)kT \log[2\cosh J/kT].   \tag{2.11}$$

Because the free energy is explicitly analytic for $T > 0$ there can be no phase transition at any finite $T > 0$. Thus the linear-chain Ising model is always in its disordered "high-temperature phase". The problem below illustrates both the absence of long-range order and the exponential decay of the short-range order.

Problem 2.3.   Show that the spin-spin correlation functions satisfy $\langle s_n s_{n+p} \rangle = \exp -|p|/\zeta$, where $\zeta(T)$ the "correlation length" is given by $\zeta(T) = 1/\log[\coth(J/kT)]$. Show that in $\lim \cdot T \to 0$, $\zeta(T) \to \exp(2J/kT)$.

Problem 2.4.   Obtain the internal energy *per* bond from F and compare with the calculated $-J\langle s_n s_{n+1}\rangle$.

This solution to this one-dimensional model dates to 1925 and provides an interesting footnote to the history of physics. Ising believed his result to be applicable in three dimensions. Because the lack of long-range order invalidated the model as a theory of ferromagnetism, far more complicated alternatives proposed by Bloch, Heisenberg, and others took over the field — for a while. Ising became a High School science teacher in New Jersey.

Almost two decades were to pass before the misapprehension was laid to rest in 2D — where Kramers and Wannier showed it *does* exhibit long-range order and spontaneous symmetry-breaking below a critical temperature T_c. It was in the 1940's that Onsager derived the exact free energy for the 2D Ising model, both below and above T_c. This discovery would usher in the contemporary era in statistical mechanics. In a later chapter we return to this problem, using the methods of quantum field theory to achieve an exact solution of this — essentially classical — problem. But the solution to the 3D model still lies — tantalizing — just beyond our grasp.

2.8. The Disordered Ising Chain

The 1D chain is readily generalized to instances in which the energy of the nth bond is $-J_n s_n s_{n+1}$ and the J_n are arbitrary. In a random chain they are distributed according to a normalized probability distribution $P(J)$. Replacing Eq. (2.11) we obtain straightaway,

$$F = -(N-1)kT \int dJ P(J) \log[2 \cosh J/kT]. \qquad (2.12)$$

Upon differentiating this expression the specific heat is obtained in the form:

$$c = C/(N-1) = k(kT)^{-2} \int dJ P(J) J^2 / \cosh^2(J/kT). \qquad (2.13)$$

We can deduce the low-temperature limiting behavior from this expression, provided $P(J)$ is approximately constant at a value $P(0)$ for small $|J|$. Then, in the range of T where kT is small compared with the typical $|J|$, we obtain

$$c = \gamma T, \quad \text{where } \gamma = k^2 P(0) \int dx \left(\frac{x}{\cosh x}\right)^2 = 1.64493 k^2 P(0). \qquad (2.14)$$

This result is just one example of what one finds in "glassy" or amorphous materials. These typically exhibit a universal law, $c \propto T$ at low temperatures — regardless of such "details" as the dispersion, dimensionality, coordination number, topology, etc.

2.9. Other Magnetic Systems in One Dimension

The Ising model is not unique in being easily solved in 1D. The statistical properties of similar other models are just as simple, provided there are no external fields coupling to the natural order parameter. "Simple" means that, as in (2.12), F takes the form:

$$F = -NkT\phi(a|T), \qquad (2.15)$$

where ϕ involves T and a set of parameters symbolized by a.

As a generic example, consider the *classical* Heisenberg model, in which vector spins are represented by classical dipoles s of fixed length s but arbitrary orientation and the total energy is the sum of all nearest-neighbor scalar bond energies $-J s_n \cdot s_{n+1}$. If we use the unit vector s_n/s to define the z-axis for the next spin, s_{n+1}, the spins decouple and the free energy is then just $N-1$ times that of a single spin, each in an effective "external" magnetic field Js. Here,

$$\phi(J|T) = \log\left\{\int d\Omega \exp-(Js^2\cos\theta)/kT\right\}.$$

Performing the indicated integration over the unit sphere, we find:

$$\phi(J|T) = \log\{4\pi(kT/Js^2)\sinh(Js^2/kT)\}. \tag{2.16}$$

As in the example of the Ising model, the very construction of this solution precludes long-range order at any finite T. Indeed, the free energy is analytic in T for all $T > 0$, and there is no phase transition.

Problem 2.5. (A) Calculate and plot the internal energy *per* site $-Js^2\langle\cos\theta\rangle_{TA}$, the specific heat, and the entropy for the Heisenberg model, as a function of T, and compare to the Ising model results. What are the limiting values as $T \to 0$ and asymptotically, as $T \to \infty$? What is the formula for the correlation function $G_p = \langle\cos\theta_n\cos\theta_{n+p}\rangle$? (B) Generalize the model and repeat the calculations for the *disordered Heisenberg chain*, in which each J_n is distributed according to a uniform probability $P(J)$, just as in Sec. 2.8.

Chapter 3

Elements of Thermodynamics

3.1.   The Scope of Thermodynamics

Thermodynamics expresses general laws governing the transformation of one kind of energy into another. Out of the huge number of dynamical variables that characterize a macroscopic system, only an irreducible few need be specified. However few they may be, these variables are not even independent but are related through an *equation of state*.

For a fluid or gas, the usual intensive variables include temperature T, pressure p, density ρ and chemical potential μ. For magnetic systems we should add the magnetic field intensity $\mathbf{B}$ to this list, and for dielectric materials the electric field strength $\mathbf{E}$. Among extensive quantities, there are the potential and kinetic energies, entropy, free energy, heat capacity, magnetization or electric polarization and more generally, particle number N and volume V. (Of course, these last are related to the intensive quantities, density ρ and specific volume v, *via* $\rho \equiv N/V = 1/\text{v}$.) Some aspects of the two-body correlation function $S(q)$ may also be relevant.

In most instances this enumeration exhausts the quantity of thermodynamic information which is either available or required. The few variables which *have* been selected are representative of a far greater number of internal variables, N', of the order of Avogadro's number N_A *per* mole. The latter include hidden dynamical variables such as the spatial coördinates, momenta, and spins of individual particles and their collective motions such as sound waves, etc. It is necessary to average over them. Consequently the *observable properties* are governed by the law of large numbers, with Gaussian probabilities subject to the CLT.

A contrasting view of thermodynamics sees it merely as the branch of applied mathematics that deals with coupled partial differential equations in several variables. That is how we shall approach it at first. Starting only

from elementary common-sense notions we derive the equations satisfied by
the free energy, the equations of state, the Maxwell relations, chain rules,
etc., in the context of simple fluids made up out of identical particles. The
rôle of internal forces shows up only in the latter parts of this chapter.

3.2.   Equations of State and Some Definitions

The equation of state constrains the smallest number of observables required
to characterize a given phase ("state") of a given substance. For over two
centuries it has been known that a simple formula relates the pressure and
temperature of any given gas to its density in thermodynamic equilibrium.
The exact formula varies from gas to gas, but theorists have labored to
place it in a universal context. Consider the "ideal gas" law, which applies
to many gases at sufficiently high T and low ρ: it is simply $p/k_B T = \rho$.
In practical terms, this means that the equilibrium state of an ideal gas is
uniquely specified by any 2 variables out of 3.

More generally the equation of state takes the form $p/kT = \rho G(\rho, T)$,[a]
with G a function to be determined experimentally or calculated with the
aid of statistical mechanics. The mere existence of such an equation of state
ensures that partial derivatives such as $\partial V/\partial p$ (related to compressibility)
and $\partial V/\partial T$ (related to thermal expansion), to pick one example, are not
independent but are mathematically related.

At fixed N one can write the generic equation of state of any gas as
$f(p, V, T) = 0$ (with $f \equiv F - F_0$ if we wish). If then we alter p by dp, V by
dV and T by dT, this quantity also changes: $f \to f + df$. The requirement
that df also vanish leads to

$$df = 0 = \left.\frac{\partial f}{\partial p}\right|_{V,T} dp + \left.\frac{\partial f}{\partial V}\right|_{p,T} dV + \left.\frac{\partial f}{\partial T}\right|_{V,p} dT. \tag{3.1}$$

The subscripts identify the variables being held constant, in accordance with
standard notation.

For more insight into (3.1) it is useful to visualize the equation of state
$f = 0$ as an hypersurface in 3-dimensional p, V, T space. By definition of
the equation of state, the fluid is in thermodynamic equilibrium *only* on this
hypersurface. The condition $df = 0$ constrains the variables to this hypersurface and ensures that all *changes occur under conditions of thermodynamic
equilibrium.*

---

[a]Where there can be no confusion we shall omit the subscript in k_B.

Setting p constant (i.e. $dp = 0$) in (3.1) we derive the first of the equations below, (3.2a). $V = $ constant and $T = $ constant are used to obtain (3.2b) and (3.2c).

$$\left.\frac{\partial V}{\partial T}\right|_p = \left(\left.\frac{\partial T}{\partial V}\right|_p\right)^{-1} = -\frac{\partial f/\partial T|_{V,p}}{\partial f/\partial V|_{p,T}} \tag{3.2a}$$

$$\left.\frac{\partial T}{\partial p}\right|_V = \left(\left.\frac{\partial p}{\partial T}\right|_V\right)^{-1} = -\frac{\partial f/\partial p|_{V,T}}{\partial f/\partial T|_{V,p}} \tag{3.2b}$$

$$\left.\frac{\partial p}{\partial V}\right|_T = \left(\left.\frac{\partial V}{\partial p}\right|_T\right)^{-1} = -\frac{\partial f/\partial V|_{p,T}}{\partial f/\partial p|_{V,T}}. \tag{3.2c}$$

The so-called *chain rule* relates the product of three derivatives:

$$\partial V/\partial T|_p \cdot \partial T/\partial p|_V \cdot \partial p/\partial V|_T = -1, \tag{3.3}$$

an identity that the reader can easily, and should, verify.

Let us expand the internal energy, $E(\mathscr{S}, V, N)$, as we did f, but without specifying that E remain constant:

$$dE = \partial E/\partial\mathscr{S}|_{V,N}d\mathscr{S} + \partial E/\partial V|_{N,\mathscr{S}}dV + \partial E/\partial N|_{\mathscr{S},V}dN. \tag{3.4}$$

All four differentials dE, $d\mathscr{S}$, dV and dN are *extensive*, therefore all three *coefficients* on the *rhs* are intensive. Identifying them with some common-place, measurable, quantities we name them *temperature*, *pressure* and *chemical potential*:

$$T = \partial E/\partial\mathscr{S}|_{V,N}, \quad p = -\partial E/\partial V|_{N,\mathscr{S}}, \quad \text{and} \quad \mu = \partial E/\partial N|_{\mathscr{S},V}. \tag{3.5}$$

These are *definitions* that supersede any qualitative notions we may have harbored concerning the "real" meaning of these primordial variables. However, it is not always convenient to use the extensive variables $\mathscr{S}$, V and N in (3.4) to determine the change in energy. Were it necessary to specify T, V and N instead, how would one reëxpress the energy? The answer is simple, even though requires defining a new entity, the *Helmholtz Free Energy* $F \equiv E - T\mathscr{S}$. Upon plugging in Eq. (3.4) we obtain,

$$dF = dE - d(T\mathscr{S}) = dE - Td\mathscr{S} - \mathscr{S}dT = -\mathscr{S}dT - pdV + \mu dN. \tag{3.6}$$

The *infinitesimals* on the *rhs* reveal the new *independent* variables, thus we write: $F = F(T, V, N)$. The change from E to F with the concomitant change in the set of independent variables constitute what is known as a *Legendre transformation*.

Two more such "free energy" quantities are constructed through additional Legendre transformations. The *Gibbs Free Energy* is defined as $G = F + pV$, therefore

$$dG = dF + pdV + Vdp = -\mathscr{S}dT + Vdp + \mu dN \,. \tag{3.7}$$

Again the infinitesimals help us identify the independent variables, T, p, and N. Thus G is $G(T, p, N)$. A third and final[b] transformation defines the *enthalpy* $\mathscr{H}(\mathscr{S}, p, N) = G + T\mathscr{S}$, i.e.

$$d\mathscr{H} = Td\mathscr{S} + Vdp + \mu dN \,, \tag{3.8}$$

for which $\mathscr{S}$, p and N are the independent variables.

Each of the free energies has its preferred applications. $\mathscr{H}$ is the simplest quantity in problems involving flow or chemical reactions, wherever a "pressure head" Δp is specified. F is easiest to calculate with the aid of statistical mechanics. As for G, a function of only a single extensive variable N, its role is paramount in the study of phase transitions.

We next want to prove a remarkable identity, $G = \mu N$. Starting with (3.4), after multiplying each extensive quantity by an arbitrary dimensionless parameter λ (e.g. $\lambda = 2$ would correspond to doubling the size of the system), we obtain: $d(\lambda E) = Td(\lambda \mathscr{S}) - pd(\lambda V) + \mu d(\lambda N)$. Treating λ and $d\lambda$ as independent variables and collecting coefficients one finds:

$$\lambda[dE - (Td\mathscr{S} - pdV + \mu dN)] = -d\lambda[E - (T\mathscr{S} - pV + \mu N)] \,. \tag{3.9}$$

By (3.4) the *lhs* vanishes. Hence so must the *rhs*. Because $d\lambda$ is arbitrary the square bracket must vanish, proving the contention:

$$[E - (T\mathscr{S} - pV + \mu N)] = F + pV - \mu N = G - \mu N = 0 \,, \quad QED.$$

It is now possible to generalize the definitions in Eq. (3.5):

$$\left.\begin{array}{l} T = \partial E/\partial \mathscr{S}|_{V,N} = \partial \mathscr{H}/\partial \mathscr{S}|_{p,N} \\[4pt] p = -\partial E/\partial V|_{N,\mathscr{S}} = -\partial F/\partial V|_{N,T} \\[4pt] \mu = \partial E/\partial N|_{\mathscr{S},V} = \partial F/\partial N|_{T,V} = \partial G/\partial N|_{T,p} = G/N \\[4pt] \mathscr{S} = \partial F/\partial T|_{V,N} = \partial G/\partial T|_{p,N} \end{array}\right\} \,. \tag{3.10}$$

The definitions of the various thermodynamic variables in (3.10) are not just pedantic. They clarify what originally had been very fuzzy ideas

---

[b]Four more Legendre transformations, consisting of subtracting μN from each of E, F, G, and H, yield only trivially different free energies.

concerning thermodynamic systems. For instance, the very *notion* of temperature was confused with that of heat until the end of the 18th Century, when it came under the scrutiny of the American physicist Benjamin Thompson (later, Count Rumford), an expatriate colonial residing in England. Further refinements are due to a French military engineer, Sadi Carnot, and an Englishman, James P. Joule. Their names remain in the core vocabulary of physical sciences and engineering.

3.3. Maxwell Relations

Let us differentiate the quantities in (3.10) once more, making use of the identity $\partial^2 f(x,y)/\partial x\partial y = \partial^2 f(x,y)/\partial y\partial x$, valid wherever f is an analytic function of its variables. For example, at constant N:

$$\left.\frac{\partial p}{\partial T}\right|_V = \frac{\partial}{\partial T}\left[-\left.\frac{\partial F}{\partial V}\right|_T\right]_V = \frac{\partial}{\partial V}\left[-\left.\frac{\partial F}{\partial T}\right|_V\right]_T = \left.\frac{\partial \mathscr{S}}{\partial V}\right|_T.$$

This establishes the first of four Maxwell relations (all of them obtained in similar fashion). They are,

$$\partial p/\partial T|_V = \partial \mathscr{S}/\partial V|_T \qquad (3.11a)$$

$$\partial T/\partial V|_{\mathscr{S}} = -\partial p/\partial \mathscr{S}|_V \qquad (3.11b)$$

$$\partial V/\partial \mathscr{S}|_p = \partial T/\partial p|_{\mathscr{S}} \qquad (3.11c)$$

$$\partial \mathscr{S}/\partial p|_T = -\partial V/\partial T|_p. \qquad (3.11d)$$

Their number can be augmented by additional degrees of freedom, such as the magnetic variables in the following example.

Problem 3.1. Prove (3.11b–d). Then extend E and the various free energies by adding $-\mathbf{B}\cdot dM$ to each. How many new Maxwell relations can one derive involving $\mathbf{B}$ and/or M and what are they?

3.4. Three Important Laws of Thermodynamics

(1) *The First Law.* Energy is Conserved.
(i) Suppose work dW is done on a closed system and "heat" dQ is simultaneously introduced into the same system, whilst carefully maintaining

thermodynamic equilibrium. The First Law states that the change in the thermodynamic function E, the internal energy, is $dE = dW + dQ$.

(ii) If the work is performed *adiabatically*, such that it preserves all classical adiabatic invariants or internal quantum numbers (i.e. in such a way that the number of accessible configurations remain constant), then $\mathscr{S}$ *remains constant* and $dE = -pdV = dW$ for mechanical work.[c] It follows that $dQ = 0$. (For work performed by an external field on a magnetic substance one has $dE = -\mathbf{B} \cdot dM$ instead, etc.)

(iii) If, on the other hand, heat is introduced through the walls with no change in volume, magnetization, nor in any other such variable, $dW = 0$ hence $dE = Td\mathscr{S}$. Thus the First Law as stated here and in (ii) above essentially restates the definitions of T in Eqs. (3.5) and (3.10) when N is constant.

(iv) If, however, $dN \neq 0$, the same equations yield $dE = \mu dN$. This expresses the fact that particles enter the system carrying energy μ.[d]

(v) Not being a relativistically invariant quantity, *energy is not conserved* in systems accelerated to near the speed of light. Therefore the First Law is modified in relativistic thermodynamics.

(2) *The Second Law:* In Equilibrium, Entropy is Maximized.

(i) In its simplest terms the Second Law states that thermodynamic equilibrium corresponds to maximal probability. Thus the approach to equilibrium has to be characterized by an entropy that increases steadily up to the asymptotic value corresponding to equilibrium. This defines an essentially *irreversible process*. In a *reversible* process with $dW = 0$ the entropy necessarily remains fixed at whatever maximal value it had.

(ii) It follows that when two closed systems are brought into thermal contact, heat flows from the hotter to the cooler until the temperatures are equal. [Proof: let $T_1 = T - t$, $T_2 = T + t$, where $|t| < T$. Invoke the First Law: $(T - t)\partial \mathscr{S}_1 + (T + t)\partial \mathscr{S}_2 = 0$, hence $\partial \mathscr{S}_1/\partial \mathscr{S}_2 = -(T+t)/(T-t)$. Expressing $\partial \mathscr{S} = \partial \mathscr{S}_1 + \partial \mathscr{S}_2$ in terms of $\partial \mathscr{S}_2$ we find $\partial \mathscr{S} = \partial \mathscr{S}_2(-2t/(T-t))$. Therefore $\partial \mathscr{S} > 0$ implies $\partial \mathscr{S}_2 < 0$. Because $\partial \mathscr{S}_2 \approx \partial \Delta Q_2/T_2$, $\Delta Q_2 < 0$. QED.]

(iii) If *particles* are permitted to flow from one container to the another at equal pressure and temperature, they do so from the higher chemical

---

[c]p = force *per* unit area and $dV = dxdA$, where x is normal to A.
[d]Special interpretations are required if μ lies in a range where there are no admissible quantum states. (Well-known examples include lightly-doped semiconductors and the "low-temperature" "s-gapped" superconductors.)

potential into the lower, until the two become equal. [Proof: along the lines of (ii). Details left to the reader.]

(iv) The entropy of a system ("A") *cannot decrease* merely by chance; any decrease must be the consequence of some physical, chemical or biological organizing force acting on A. Then, suppose A to be entirely embedded in a larger, *closed*, system B. Even if the entropy of A is decreasing the *total* entropy of B must *increase*, as shown in example (ii) above. The exothermic combination of two species to form a third, e.g. $\alpha + \beta \to \gamma$, provides a less obvious example. The species' $\mathscr{S}$ obviously decreases but once the entropy of the released radiation ("heat") is taken into account the *total* $\mathscr{S}$ again increases.

(3) *The Third Law* (Nernst's 1905 Theorem): $\mathscr{S}/N \to 0$ as $T \to 0$

(i) From basic assumptions we know that, in equilibrium, entropy increases with total energy; thus it must be a minimum in the state of lowest energy, i.e. at $T = 0$. The Third Law stipulates that the entropy *per* particle is not just a minimum, but that it actually *vanishes* in the ground state. This can be proved rigorously for bosons, as their ground state wavefunction is nodeless (regardless of interactions), hence unique. This Law seems to hold in other contexts as well, so the result may be more general than the derivation. Known counter-examples have invariably proved to be artificial, mathematical, models devoid of fluctuations.

(ii) As a consequence of Nernst's theorem, in the lim $\cdot T \to 0$ the specific heat, $c(T) \equiv T\partial s(T)/\partial T$, also vanishes. The proof is almost trivial:

$$s(T) \equiv \mathscr{S}(T)/N = \int_0^T dT' \frac{c(T')}{T'}$$

cannot vanish at $T = 0$ unless c does. The vanishing of $c(T)$ at $T = 0$ is sometimes (incorrectly) referred to as the Third Law although that alternative formulation could be paradoxical (see below).

Problem 3.2. An "electron gas" (i.e. metal) has low-temperature heat capacity $c(T) = \gamma T \to 0$ as $T \to 0$. The low-temperature heat capacity of acoustic phonons is $c(T) \propto T^3$, also vanishing in the lim $\cdot T \to 0$. *Show* both examples are compatible with the Third Law. By way of contrast, consider an hypothetical system A, in which $c_A(T) = c_0/\log(1 + T_0/T)$, with c_0 and T_0 fixed parameters. At $T = 0$, $c_A(T)$ *also* vanishes. Nevertheless, A violates the Third Law as given above. *Show this.*

Historically the Third Law antedated quantum mechanics although its significance only became clear after the symmetries of quantum mechanical many-body ground states were revealed. In this text we use it to obtain a simple identity in the theory of superconductivity. Although occasionally one may find this law useful, Simon[e] surely overstated the case in 1930 by calling it "...the greatest advance in thermodynamics since van der Waals' time..." even though this assessment concurred with Nernst's high opinion of his own discovery. Dugdale[f] reports that

> ... Nernst was very proud of his achievement in formulating what he referred to as "my law". He noted there were three people associated with the discovery of the First Law, two with the Second and only one with the Third. From this he deduced that there could be no further such Laws.[f]

But not all writers agreed on the importance of the Third Law; in fact, Fowler and Stern called it "irrelevant and useless".[f] Nor is it destined to be the final word: there have already been formulations of a "Fourth Law", in a sequence that will not likely end soon.

3.5. The Second Derivatives of the Free Energy

With few exceptions, the principal *intrinsic* characteristics of matter have been found to be functions that can be expressed as a second derivative of some free energy, such as the previously discussed *heat capacity* and *magnetic susceptibility*. Additional quantities of this type are *compressibility* and *dielectric susceptibility*.[g] We next derive some relations and inequalities governing such functions. For example, define $C_{\rm v} = T\partial\mathscr{S}/\partial T|_{\rm v}$. Then,

$$C_{\rm v} = -\left.\frac{T\partial^2 F}{\partial T^2}\right|_{\rm v}. \quad \text{Similarly,} \quad C_p = -\left.\frac{T\partial^2 G}{\partial T^2}\right|_p. \tag{3.12}$$

Fix N and expand: $d\mathscr{S} = \partial\mathscr{S}/\partial p|_T dp + \partial\mathscr{S}/\partial T|_p dT$. After multiplying by T/dT and specifying $V = $ constant, we find $Td\mathscr{S}/dT|_{\rm v} = $

---

[e]F. Simon, *Ergeb. Exakt. Naturwiss.* **9**, 222 (1930).

[f]See discussion in J. S. Dugdale, *Entropy and Its Physical Meaning*, Taylor & Francis, London, 1996, p. 145.

[g]At finite frequency and wavevector this generalizes to the complex dielectric function $\varepsilon(\omega, q) = \varepsilon_1(\omega, q) + i\varepsilon_2(\omega, q)$. The imaginary part of ε relates to electrical conductivity, Joule heating, etc.

$T\partial\mathscr{S}/\partial p|_T dp/dT|_v + T\partial\mathscr{S}/\partial T|_p$, relating C_v to C_p. The Maxwell relation (3.11d) eliminates $\partial\mathscr{S}/\partial p|_T$ in favor of $-\partial V/dT|_p$; the chain rule (3.3) eliminates the latter. Then,

$$C_p = VT\kappa_T \left(\left.\frac{\partial p}{\partial T}\right|_V\right)^2 + C_v. \tag{3.13}$$

In the above we used the *isothermal* compressibility, a material property that is itself a second derivative of Gibbs' free energy:

$$\kappa_T \equiv -\frac{1}{V}\left.\frac{\partial V}{\partial p}\right|_T$$

$$= -\frac{1}{V}\left.\frac{\partial^2 G}{\partial p^2}\right|_T. \quad \left(\text{Similarly}: \kappa_{\mathscr{S}} \equiv -\frac{1}{V}\left.\frac{\partial V}{\partial p}\right|_{\mathscr{S}} = -\frac{1}{V}\left.\frac{\partial^2 E}{\partial p^2}\right|_{\mathscr{S}}\right) \tag{3.14}$$

(Note the $-$ signs). $\kappa_{\mathscr{S}}$ is the *adiabatic* (or *isentropic*) compressibility. It is a given that compressibility is *non-negative* (application of pressure causes compression, not dilation)! Thus Eq. (3.13) serves to establish a general inequality: $C_p > C_v$.

In *magnetic systems* one defines *isothermal* and *adiabatic* magnetic susceptibilities as:

$$\chi_T = \left.\frac{\partial M}{\partial B}\right|_T \quad \text{and} \quad \chi_{\mathscr{S}} = \left.\frac{\partial M}{\partial B}\right|_{\mathscr{S}}, \text{ respectively.} \tag{3.15}$$

The heat capacities C_B and C_M are then related by,

$$C_B = C_M + T(\chi_T)^{-1}\left[\left.\frac{\partial M}{\partial T}\right|_B\right]^2. \tag{3.16}$$

But unlike the compressibility, which is always positive, χ_T can either be positive (in *paramagnetic* or *ferromagnetic* materials) or negative (in *diamagnetic* materials, including all superconductors). Therefore the sense of the inequality $C_B \geq C_M$ depends on the sign of χ_T.

3.6. Phase Diagrams for the van der Waals Gas

A plot of the equation of state in the neighborhood of the ferromagnetic-paramagnetic phase transition, as in Fig. 2.1, or of the liquid-vapor phase transition, yields valuable information.

Consider the liquid-vapor phase transition. Certainly the ideal-gas equation of state, $p = NkT/V = kT/v$, does not admit a condensed phase; its isotherms (plots of p versus v at constant T) march in parallel formation, with nary a kink.

The first systematic theory of a non-ideal gas was that of the Dutch scientist, Johannes D. van der Waals [1837–1923], who derived the following equation of state entirely on empirical grounds:

$$p = \frac{kT}{v - b} - a \left(\frac{1}{v}\right)^2 . \tag{3.17}$$

The parameter b measures the excluded volume *per* particle, and a the intermolecular attraction. The measured b turned out to depend on v and T for reasons which escaped van der Waals, but which we can now understand on the basis of the *virial expansion* (*vide infra*). Nevertheless, on the basis of this equation of state he was able to explain the existence of a critical point (p_c, v_c, T_c) for the vapor-liquid phase transition (this had already been experimentally identified by 1869) and to derive a *Law of Corresponding States* in 1880.[h] In his Nobel prize acceptance speech, van der Waals saw the qualitative agreement of his theory with experiment as a major victory for the "atomistic" theory of matter — stressing that this view had still remained controversial at the turn of the Century!

The critical point can be derived from (3.17) by identifying the equation as a cubic in v and finding the point at which its 3 roots coincide. One solves 3 equations for the 3 unknowns: $dp/dv = 0$ (i.e. $\kappa_T \rightarrow \infty$), $d^2p/dv^2 = 0$ together with Eq. (3.17) itself. As a result, Eq. (3.17) can be replaced by a generic,

$$\tilde{p} = \frac{8\tilde{t}}{3\tilde{v} - 1} - \frac{3}{\tilde{v}^2} , \tag{3.18}$$

where $\tilde{p} = p/p_c$, $\tilde{v} = v/v_c$, and $\tilde{t} \equiv T/T_c$, and $p_c = \frac{a}{27b^2}$, $v_c = 3b$, $kT_c = \frac{81}{27b}$.

This reformulation in term of dimensionless and presumably universal variables renders van der Waals' Law of Corresponding States explicit.

Neat as it appears, Eq. (3.18) is not entirely satisfactory. Plotting it one can identify not just two distinct phases (gas and liquid) but also regions of negative compressibility and of negative pressure — both proscribed in thermal equilibrium. The necessary "fix" to the coexistence region is provided by Maxwell's construction discussed below, which achieves one of the requirements of the Second Law, i.e. that $\mu_{vapor} = \mu_{liquid}$ whenever the two

---

[h] According to this Law, all gases satisfy a universal equation of state once their variables are expressed in dimensionless form.

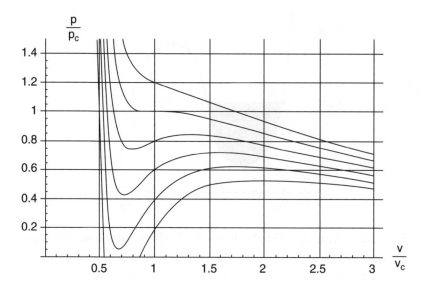

Fig. 3.1.   Isotherms in van der Waals' universal equation of state.

$\tilde{p}$ the pressure in dimensionless units) versus $\tilde{v}$ (the volume per molecule in dimensionless units) at various (dimensionless) temperatures, as in Eq. (3.18). *From bottom curve to top:* $T/T_c = 0.80, 0.85, 0.90, 0.95, 1.00, 1.05$.

fluids coexist. However, Maxwell's construction is paradoxical insofar as it makes use of van der Waals' isotherm $p(v)$ in the *unphysical region*.

Experimentally, laws of Corresponding States somewhat different from (3.18) have been found to hold in a variety of different fluids, although *details* such as *power laws* disagree with (3.18). Indeed, the search for a better formula has spurred an entire field of investigation concerned with "critical point phenomena".[i]

Clearly the plotted isotherms are unphysical wherever p is negative, and also wherever their slope is positive (i.e. where the compressibility is

---

[i]Near second-order phase transitions length scales such as the interparticle separation or the lattice parameter becomes irrelevant as fluctuations occur at all wavelengths. This allows to construct correlation lengths based on small quantities such at $T - T_c$ or B (external field) and to express all thermal properties in terms of these. This "scaling" "critical theory" attributable in large part to early work by M. E. Fisher and by L. P. Kadanoff is the subject of H. E. Stanley's *Introduction to Critical Phenomena*, Oxford, 1971. Later, *ca.* 1984, it was discovered that in 2D the continuum hypothesis is well matched to mathematical concepts of *conformal invariance*. This identification allows to solve practically any 2D physical model near its phase transition. Many relevant papers are to be found in C. Itzykson, H. Saleur and J. Zuber's massive reprint volume, *Conformal Invariance and Applications to Statistical Mechanics*, World Scientific, 1988.

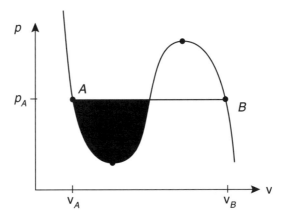

Fig. 3.2.   van der Waals' loop, and Maxwell's construction.

Along a given isotherm, with increasing v the pressure first decreases to A in accordance with Eq. (3.17), then stays constant (at the value p_A determined by Eq. (3.19)) while the ratio of liquid-to-vapor diminishes until the pure vapor phase is reached at B. Thereafter $p(v, T)$ satisfies Eq. (3.17) once again. The black loop has the same area as the white, hence Maxwell's construction Eq. (3.19) applies. The fraction ρ_A of liquid is a variable decreasing from 1 to 0 in the coëxistence region, along the straight line joining A to B.

The *entire* curve between points A and B is unphysical in thermodynamic equilibrium — not merely the portions (between the two dots) that exhibit negative compressibility. The weakness of Maxwell's construction is that it uses the solution of Eq. (3.17) to position the points A and B *precisely in the range where the equation is unphysical, hence unreliable.*

negative). Maxwell's construction, shown in Fig. 3.2, connects the point $A(T)$ on the liquid side (small value of v) to the point $B(T)$ on the vapor side (large v), by an horizontal line at $p_A \equiv p(v_A, T) = p(v_B, T)$. The value of p_A is determined by the nonlocal condition that the work from A to B *along* the isobar equals the integrated work along the van der Waals loop. This fixes $\mu_A = \mu_B = \mu$, allowing the free interchange of molecules between the liquid (at A) and the vapor (at B) at constant p, T. *Quantitatively:*

$$p_A \cdot (v_B - v_A) = \int_A^B dv p(v, T) \,. \tag{3.19}$$

The integrand $p(v, T)$ is that given in Eq. (3.17). Figure 3.2 illustrates the solution of Eq. (3.19), $p_c(T)$ and $v_{A,B}(T)$, for a typical isotherm below T_c.

The locus of points $A(T)$ and $B(T)$ defines a *coëxistence curve* in the p, v plane, in the shape of a distorted inverted parabola with its apex at

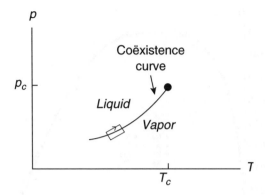

Fig. 3.3.   **Coexistence curve in p, T plane.**

Crossing the coëxistence curve (left to right) at a point p, T changes v from v_A to v_B discontinuously. This denotes a *first-order* phase transition characterized by a *latent heat*, $L(T)$ which decreases with increasing T to $L(T_c) = 0$. At T_c precisely, $v_A = v_B \equiv v_c$ and the transition is *second-order*. For $T > T_c$ the equation of state is continuous (no transition) and the two phases are indistinguishable, merged into a single fluid phase. The box illustrates the construction in the Clausius–Clapeyron relation.

the point (v_c, p_c).[j] The agreement with experiment is *qualitatively* good albeit *quantitatively* defective: the measured curve is flatter than a parabola at its apex. In Fig. 3.4 we exhibit experimental results from a variety of fluids. While displaying universal behavior predicted by van der Waals in his Law of Corresponding States, the curve does not assume the parabolic shape required by his theory. It is better fitted to a cubic form.

Above the critical point $(T > T_c)$ where the isotherms decrease monotonically with v, the vapor can no longer be objectively distinguished from the liquid. This is best seen in the schematic plot of $p(v_A(T), T)$ versus T in Fig. 3.3.

3.7.   Clausius–Clapeyron Equation

First-order phase transitions such as those just documented are characterized by a discontinuity in a *first derivative* of some free energy such as the entropy, internal energy, or density. Given the coëxistence of the two phases, the Second Law generalizes Maxwell's construction by guaranteeing that the

---

[j]Or, in the dimensionless coordinates of Fig. 3.1, at the point (1,1).

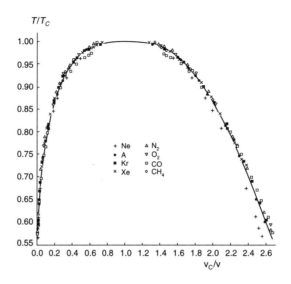

Fig. 3.4.[k] Temperature (units T_c) versus density (v_c/v).

When plotted in dimensionless units, the coëxistence curves (the locus of points A and B for all temperatures $T < T_c$ in Fig. 3.2) in the $p, 1/v$ diagram of 8 disparate fluids, all lie on, or close to, a single universal curve: a sort of "Law of Corresponding States". Although van der Waals' equation together with Maxwell's construction predicts a parabolic T versus $1/v$ dependence near the critical point at $(1,1)$, the experimental results are more nearly cubic (solid curve).

Gibbs free energy *per* particle, μ, is the same in both phases at given T and p.[l]

The Clausius-Clapeyron equation is based on this simple observation. It relates the difference in specific volume $\Delta v = v_B - v_A$ between the two phases to the difference in the specific entropies (entropy *per* particle) s. Consider *two* distinct infinitesimal displacements along the critical curve in Fig. 3.3, starting from a point (T, p). The first displacement, *immediately above* the curve, is in the liquid phase and is parametrized by dT and s and dp and v. The other, *immediately below*, lies in the vapor phase and is parametrized by dT and s' and dp and v'. Because the $\mu's$ remain equal, $d\mu' = -s'dT + v'dp = -sdT + vdp = d\mu$. Collecting coefficients of dp and dT and defining

$$L(T) \equiv T(s' - s) = T\Delta s$$

[k] Adapted from E. A. Guggenheim, *J. Chem. Phys.* **13**, 253 (1945).
[l] See statement of Second Law, part (iii).

the *latent heat* for the transition, we obtain the Clausius-Clapeyron equation,
in truth an identity,

$$\frac{dp}{dT} = \frac{L}{T\Delta v}(\text{both } L \text{ and } \Delta v \text{ being known functions of } T). \qquad (3.20)$$

The *lhs* is obtained from experiment (Fig. 3.3), as is Δv (Fig. 3.4). Thus
(3.20) is used to measure s experimentally. If dp/dT remains finite at T_c this
formula also links two apparently independent phenomena: the vanishing of
L at the critical point T_c and the vanishing of Δv at that same point.

But this last applies only to a phase transition that connects two *fluid*
phases of one and the *same* material. If the condensed phase were a solid
there could be no critical point at finite T. (We shall explain this important
distinction between the solid state and the liquid by invoking the long-range
order associated with the former.) Nor would the phase diagram be quite so
simple if there were several distinct fluids in solution.

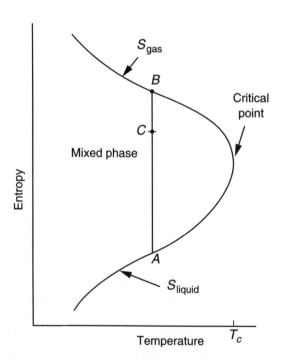

Fig. 3.5.   Entropy versus temperature along coexistence curve.
The latent heat can be obtained using the formula $L(T) = T(\mathscr{S}_{\text{gas}} - \mathscr{S}_{\text{liquid}})$ up
to the critical temperature. Note that L must vanish at T_c if the curves A and B
meet without discontinuity, as in the drawing. The gas fraction in the mixed phase
region between the two curves is approximately $(C - A)/(B - A)$.

The source of latent heat in the phase transition governed by the Clausius–Clapeyron Eq. (3.20) is exhibited graphically in Fig. 3.5 above.

3.8.   Phase Transitions

Even though the latent heat vanishes at the critical point, as illustrated on the preceding page, the isothermal compressibility diverges, as is apparent from one of the equations $dp/dv = 0$ that helps define T_c.

Thus a phase transition which is first order across the coëxistence curve at temperatures $T < T_c$ become *second-order* at its end-point T_c, before being extinguished altogether. We have noted that second-order phase transitions are the result of a discontinuity (or a singularity) in *any* second derivative of the free energy; the compressibility is just such a quantity. In some instances, it is a third-order derivative that is discontinuous.

But in all such cases, or under even more obscure circumstances where the discontinuity is fourth order or higher, or in instances where all derivatives are continuous but the free energy exhibits an essential singularity,[m] the conventional nomenclature remains the same: *any* phase transition that is *not* first-order is deemed, by default, to be a *second-order* phase transition.

If the phase on one side of a phase boundary has long-range order and that on the other side does not, there is no critical point. The explanation is this: in Fig. 3.3 we see it is possible to proceed continuously from the vapor phase to the liquid by circling about the critical point without crossing the critical line nor any hypothetical curve at which long-range order might set in. Thus, if a critical fluid-solid critical point were to exist, it would have to be at $T \to \infty$ to forbid such trajectories. Interestingly, a solid *can* coëxist with its liquid and vapor both, at a "triple point" which marks the intersection of the two coëxistence curves.

The familiar example of H_2O illustrates this, in Fig. 3.6 next.

This figure perforce omits many interesting details, such as the many different phases of ice and the solid-solid phase transitions which separate them. On this scale one can barely observe the *negative slope* of the *solid-liquid* coëxistence curve near the triple point, due to the lower density of ice relative to water. There exist microscopic theories for this physical attribute, uniquely responsible for the winter sports of ice skating and skiing.

---

[m]E.g.: $F = F_0 + F_1 \exp(-|\Delta|/|T - T_c|)$, and all derivatives of F, are continuous at T_c in the case of the two-dimensional "plane-rotator", but there is nevertheless a phase transition.

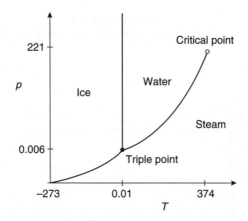

Fig. 3.6. Schematic phase diagram for H$_2$O.

Pressure (*bars*) versus Temperature (°C).

Distinct phases of a given molecular species may differ in their density, texture, compressibility, susceptibility, etc., and will generally also differ in their *order parameters* including X-Ray structure factors, magnetization, electric polarization, and the like. Phase diagrams of compound substances present a more complex set of issues. In the solid state, the various ordered and disordered alloys compete while in the liquid phase one finds either intimate mixes or phase separations such as oil on water. Whatever the case, the appropriate generalization of the total Gibbs' free energy in any of these applications is always just $G = \sum \mu_j N_j$. The individual chemical potentials given by $\mu_j = \partial G / \partial N_j |_{T,p}$ are continuous across all phase boundaries. Armed just with this observation (plus elementary algebra and some uncommonly good "common sense"), Gibbs devised the following inequality governing the coëxistence of r phases in a mix of n species at some fixed, given, p and T.

Gibbs' Phase Rule: $r \geq n + 2$. The deficit, $f = n + 2 - r$, equals the total number of independent variables that have to be specified when the r phases are in thermodynamic equilibrium.

- *Example*: in the case of pure H$_2$O $n = 1$. Therefore the maximum number of species r (water, vapor, ice) is ≥ 3. No more than 3 phases can coëxist — *vide* Fig. 3.6 showing the unique "triple point" with $f = 0$ (hence requiring no extra variable). Where just 2 phases coëxist, as in the water/vapor coëxistence region, the deficit $f = 1$ and precisely 1 free variable is required. It may be chosen as $\rho(\mathrm{v})$, ranging from ρ_A to ρ_B along the isobar from A to B in Fig. 3.2 or along the isotherm in Fig. 3.5.

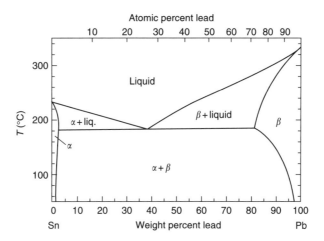

Fig. 3.7. Phase diagram for mixtures of tin and lead.[n]

Temperature versus composition for $n = 2$, $r = 4$. The α phase refers to a Sn structure with Pb impurities, β to the equivalent Pb structure, and $\alpha + \beta$ to the solid-state alloy of the two. Starting from the liquid mixture at high temperature, the condensation occurs initially for one or the other constituent (e.g. to $\alpha +$ liquid mixture on the left). Coëxistence regions are lines, except for the one point (where $r = 4$, hence $f = 0$) where both constituents condense simultaneously, called the "eutectic" *point*.

Far more complicated phase diagrams (such as the typical one of a binary alloy illustrated above) are simplified by the application of Gibbs' rule. The proof of it now follows.

Proof. Start by labeling the phases by $\alpha = 1, 2, \ldots, r$ and defining the density of the jth species in the αth phase as $\rho_j(\alpha)$. Each μ_j in a given phase α is a function of p, T, and of the $(n - 1)$ ratios of densities: $\zeta_2(\alpha) \equiv \rho_2(\alpha)/\rho_1(\alpha)$, $\rho_3(\alpha)/\rho_1(\alpha), \ldots, \rho_n(\alpha)/\rho_1(\alpha)$. This makes for a total $2 + r \times (n - 1)$ independent variables. The chemical potential for each species is constant, i.e. $\mu_1(1) = \mu_1(2) = \cdots = \mu_1(r)$, $\mu_2(1) = \mu_2(2) = \cdots = \mu_2(r)$, etc. Each of these $n \times (r - 1)$ equations involves all the independent variables. ☐

For a common solution to exist, the number of equations cannot exceed the number of variables. Hence $2 + r \times (n - 1) \geq n \times (r - 1)$. Rearranging the terms, $r \leq n + 2$, *QED*.

[n]From T. D. Massalski, Ed., *Binary Alloy Phase Diagrams*, ASM International, Materials Park, Ohio, 1990.

3.9.   The Carnot Cycle

We examined the non-ideal gas at constant T along an isotherm and at constant p along an isobar (known to earlier generations as an *isopiestic*). We now find it necessary to examine the trajectory at constant entropy along an *adiabatic* (a.k.a. *isentropic*). The traditional picture is that of a piston quasi-statically compressing a fluid; work is done on the working substance, while heat flows neither in nor out.

For an ideal, collisionless, monatomic, gas it is easy to show[o] that the internal energy $NkT = K/V^{2/3}$, where K is a constant. Hence, the equation of state yields $pV^{5/3} = K$ on the adiabatic trajectories, curves which are steeper than the isotherms and therefore intersect the latter. Typical adiabats and isotherms are sketched in Fig. 3.8. The closed cycle of two adiabats and two isotherms drawn in this figure illustrates the Carnot cycle — whether operated as an engine (driven clockwise) or as a refrigerator (counterclockwise).

In reality, engines and refrigerators as ordinarily constructed are subject to complex conditions — neither isothermal nor adiabatic, and are generally not operated in thermodynamic equilibrium — hence they are not amenable to any simple analysis. The working substance is often far from the ideal gas and a quasi-static operation may be out of the question. For example, the familiar internal combustion engine has taken a century to perfect simply because it functions so far out of equilibrium that it is not subject to any of the simplifications of an equation of state. Although widely studied, this particular cycle remains inherently inefficient and to this day the number of possibilities for its improvement remains endless.

In comparison with such everyday machines the hypothetical Carnot cycle is unique in its simplicity, in its efficiency, and in the way it illuminates the Second Law. Carnot devised his reversible virtual engine in 1824. Its efficiency[p] η cannot be surpassed by any real motor (cf. Problem 3.3 below). Moreover, this is a machine that can be operated as a motor (heat in, work out) or in reverse, as a refrigerator (work in, heat out). We illustrate the former mode; the reader is invited to work out the latter.

With reference to the Figure, each cycle consists of the four trajectories $A \to B, \ldots, D \to A$, in sequence. In the first leg, a piston adiabatically

---

[o]With p_x the momentum of a gas particle in the x direction, $p_x V^{1/d}$ is an adiabatic invariant in d dimensions. Therefore the particles' kinetic energy $\propto 1/V^{2/d}$. This quantity is proportional to T, hence to pV. QED.
[p]η = work output ÷ heat input.

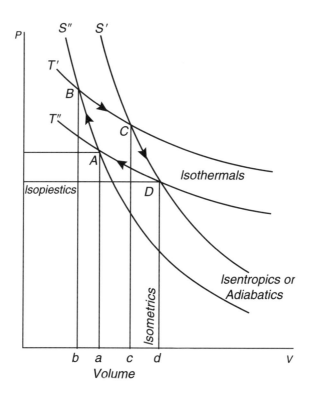

Fig. 3.8.   Schematic phase diagram of a working fluid illustrating the Carnot Engine [as adapted from 1911 Encyclopedia Britannica].
Showing curves of constant temperature (*isothermals*) and constant entropy (*adiabatics* or *isentropics*), lines of constant pressure (*isopiestics*) and constant volume (*isometrics*). The work delivered by the engine equals the area of the figure ABCD, traversed in the sense of the arrows. $T' > T''$ and $\mathscr{S}' > \mathscr{S}''$. In the refrigerator mode the sense of the arrows is reversed.

compresses the fluid from a volume V_a and temperature T'' to a smaller volume V_b at a higher temperature T'. By definition, $\Delta\mathscr{S}_{A,B} \equiv 0$. The second leg is an isothermal expansion to V_c. The change in entropy is $\mathscr{S}' - \mathscr{S}'' = \Delta\mathscr{S}_{B,C} = Q_{B,C}/T'$, where Q is the heat supplied to the working substance. The third leg is an adiabatic expansion ($\Delta\mathscr{S}_{C,D} \equiv 0$) in the course of which the temperature reverts to T'' and the volume to V_D. The final leg is an isothermal compression back to A, with $\Delta\mathscr{S}_{D,A} = -Q_{D,A}/T''$. The cycle repeats for as long as the energy source $Q_{B,C}$ is available at the temperature T'.

Because $\mathscr{S}$ is a *function of state* the sum of the changes $\Delta\mathscr{S}$ around the cycle add to zero. By conservation of energy (i.e. the First Law) the work

$W_{ABCD} = Q_{B,C} - Q_{D,A}$. Solving two equations in the two unknowns one obtains for the efficiency:[p]

$$\eta = 1 - \frac{Q_{D,A}}{Q_{B,C}} = 1 - \frac{T''}{T'}. \tag{3.21}$$

Were the process not quasi-static an *extra* $\Delta\mathscr{S}$ would be created in each cycle. If we write it in the form $\Delta\mathscr{S}_{extra} = xQ_{B,C}/T'$, with x measuring the irreversibility, the net work done is $W_{ABCD} = (1 - x)Q_{B,C} - Q_{D,A}$ and η in (3.21) is thereby reduced by x. In fact, (3.21) is the maximum efficiency available to *any* device — not just Carnot's — run between temperatures T'' and T'. The proof:

An hypothetical engine, denoted F, assumed to have efficiency exceeding η, is coupled to Carnot's cycle, R, run in refrigerator mode at the ambient temperature T''. The work delivered by F, fueled by the heat output of R at T', exceeds the work actually needed to run the refrigerator and is stored. Such a "perpetual motion" machine effectively converts ambient heat into stored energy in each cycle, thereby violating the Second Law.[q]

Problem 3.3.   Work out the details of the proof, showing that even if the First Law were satisfied by the above R–F couple the Second Law would be violated.

3.10.   Superconductivity

Some aspects of superconductors are understood by means of purely thermodynamic arguments, without benefit of microscopic theory. For example, consider features shared by a number of "conventional" superconductors[r] such as lead, mercury, niobium, etc.: as the temperature is lowered below a critical value T_c ranging from 0 to 25 K, depending on the material, a jump in the specific heat signals the onset of perfect conductivity. The extreme diamagnetism of such "type I low-T_c superconductors" serves to exclude magnetic flux at all temperatures below T_c. However, once the external field exceeds a critical value Bc it *does* penetrate the material and the normal

---

[q]For this reason the US patent office has, since the mid-19th Century, required a "working model" (!) before processing purported perpetual motion inventions.
[r]The novel *high temperature superconductors* based on CuO_2 layers have a distinct set of properties and of "corresponding states".

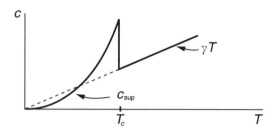

Fig. 3.9.   Heat capacity in a conventional superconductor.
The linear specific heat indicates an ordinary metallic phase and the exponential
curve a gapped superconductor. The jump in c_{sup} occurs at T_c, precisely at the
threshhold for superconducting behavior.

metallic state is restored. It is found that the ratio $B_c(T)/B_c(0)$ is a function
only of T/T_c, a sort of "law of corresponding states".

In the figure above, the measured heat capacity $c_s(T)$ of a superconductor
(at temperatures $T \leq T_c$) is compared with that of the normal metal, $c_0(T) =$
γT (at temperatures $T > T_c$, continued by a dashed line to $T = 0$). The
parameter γ is obtained experimentally.

Because the normal-to-superconducting phase transition in zero field is
second-order the entropy of both phases is continuous through the phase
transition. According to Nernst's theorem the entropy of either phase
vanishes at $T = 0$. Thus, equating the integral over c/T from $T = 0$ to
T_c in each phase leads to $\mathscr{S}_s(T_c^-) = \mathscr{S}_n(T_c^+)$:

$$\int_0^{T_c} dT \frac{c_s(T)}{T} = \gamma T_c. \tag{3.22}$$

To satisfy this equation the *excess* in the specific heat of the superconductor
just below T_c relative to that of the normal metal, γT, has to be compensated
by a *reduction* in specific heat relative to γT, over the lower temperature
range. Figure 3.9 illustrates this behavior. Equation (3.22), a sort of Maxwell
construction, is confirmed experimentally and affirmed by the microscopic
BCS theory.[s]

A second thermodynamic identity connects the critical field B_c to the
specific heat anomaly in zero field *via* the magnetic analogue of the Clausius–
Clapeyron relation. The latent heat of the superconducting-to-normal phase
transition (in which the magnetic field penetrates the metal, destroying the

[s]J. Bardeen, L. N. Cooper and J. R. Schrieffer, *Phys. Rev.* **106**, 162 (1957), and **108**, 1175
(1957).

superconducting phase) in an homogeneous field $B_c(T)$ is precisely $L = -T\partial\left(\frac{VB_c^2}{8\pi}\right)/\partial T$. Thus,

$$\Delta c \equiv c_{\text{sup}} - \gamma T = -T\frac{\partial}{\partial T}\left(\frac{L}{VT}\right) = \frac{T}{8\pi}\frac{\partial^2}{\partial T^2}(B_c^2)$$

$$= \frac{T}{4\pi}\left[\left(\frac{\partial B_c}{\partial T}\right)^2 + B_c\left(\frac{\partial^2 B_c}{\partial T^2}\right)\right]. \tag{3.23}$$

At T_c, the critical field B_c vanishes. Thus, only the first term contributes and the discontinuity is precisely equal to $\Delta c|_{T_c} = \frac{T_c}{4\pi}(\partial B_c/\partial T|_{T_c})^2$. This thermodynamic identity, too, is in excellent accord *both* with experiment and with predictions of the microscopic "BCS" theory of Bardeen, Cooper and Schrieffer.[s]

Chapter 4

Statistical Mechanics

4.1.   The Formalism — and a False Start

Like the ancient Japanese game of Go, *axiomatic* statistical mechanics has
few rules, all easy to remember. The art is entirely in the implementation. A
single formula, at the very core of the theory, defines the *Partition Function*
Z,[a] an un-normalized sum in which every allowed configuration is assigned
a probability proportional to its Boltzmann factor:[b]

$$Z = Tr\{e^{-\beta H}\}. \tag{4.1}$$

Here $\beta = 1/kT$ and Tr (the abbreviation of "trace") means: sum over all
accessible "states" (or configurations).

   Although the same definition applies whether the particles satisfy classical
dynamics or quantum mechanics, the implementation differs in the two cases.
The classical non-relativistic Hamiltonian $H\{q_j, p_j\}$ is a scalar function of a
set of $3N$ dynamical coördinates q_j and their conjugate momenta p_j. Due to
the constraints of classical dynamics H is a constant of the motion, the value
of which, E, is denoted the "energy". The energy is continuously distributed
and, in some instances, labeled by the other constants of the motion such as
momentum, angular momentum, etc. "Trace" consists of an integral over all
q_j, p_j.

   In quantum theory H is an operator; its eigenvalue is the energy E,
generally labeled by a set of quantum numbers. The energy can be discrete
(i.e. "quantized"), or, as in classical dynamics, in a continuum. "Trace" here
consists of a sum over all state-averages.

---

[a]"Z" for "Zustandsumme".
[b]In quantum parlance $Z^{-1}\exp-(\beta H)$ is designated the "density matrix".

The normalized quantity $\{Z^{-1}\exp-\beta H\}$, frequently denoted the *density matrix*, is used in the calculation of thermodynamic averages such as the internal energy $E = \langle H \rangle_{TA}$. On the other hand, Z is itself postulated to be related to a quantity F, $Z \equiv \exp-\beta F$. Soon we shall identify F thermodynamically.

Applying these relations immediately yields the following:

$$E = Z^{-1}Tr\{He^{-\beta H}\} = -\frac{\partial \log Z}{\partial \beta} = \frac{\partial(\beta F)}{\partial \beta} \qquad (4.2)$$

which is just the thermodynamic identity first derived on p. 21 on the basis of the "stationarity" of F. Combining F and E we also extract the entropy: $F - E \equiv -T\mathscr{S} = F - \partial(\beta F)/\partial\beta \equiv T\partial F/\partial T$, i.e. $\mathscr{S} = -\partial F/\partial T$, consistent with the definitions in Chapter 2.

Continuing in the same vein we differentiate the first two expressions in (4.2) to find the heat capacity $C = dE/dT$ given by:

$$C = \frac{1}{kT^2}\langle(H - \langle H\rangle_{TA})^2\rangle_{TA}, \qquad (4.3)$$

i.e. the derivative of the average energy is proportional to the variance or "noise" in the distribution of energies. Equation (4.3) proves the heat capacity to be positive definite, as previously claimed.

Frequently one probes a system by some potential H' to obtain its response characteristics. In many cases (but not all!) this perturbation affects the Hamiltonian linearly: $H \to H - gH'$. Then the system's response $\langle H'\rangle_{TA}$ is,

$$\langle H'\rangle_{TA} = \frac{Tr\{H'e^{-\beta(H-gH')}\}}{Tr\{e^{-\beta(H-gH')}\}} = kT\frac{\partial \log Z}{\partial g} = -\frac{\partial F}{\partial g}. \qquad (4.4)$$

Like Eq. (2.6), this has the solution, $F = F_0 - \int_0^g dg\langle H'\rangle_{TA}(g)$.

Differentiating (4.4) once again *w.r.* to g, we find for $-\partial^2 F/\partial g^2$:

$$\chi = \beta\langle(H' - \langle H'\rangle_{TA})^2\rangle_{TA}, \qquad (4.5)$$

similar to the formula for heat capacity. In the lim $\cdot\, g \to 0$, the generalized susceptibility $\chi_0 \equiv \partial\langle H'\rangle_{TA}/\partial g$ is one of the characteristic physical properties of the system. (It belongs to the class of "second derivatives" introduced in Chapter 3.)

Equation (4.5) expresses, and indeed proves, the following theorem:

Concavity Theorem: Whenever $H = H_0 - gH'$, $\partial^2 F/\partial g^2 < 0$ $\forall g$.

We start actual calculations with *classical* statistical mechanics. Define the "trace" over all coördinates and momenta as:

$$Z_0 = (2\pi\hbar)^{-3N} \prod_{j=1}^{N} \int d^3q_i \int d^3p_j \exp\{-\beta H\}\,.$$

The integration is over each of N point particles in 3D. The factor $(2\pi\hbar)^{-3}$, introduced into statistical mechanics well before the discovery of quantum theory, defines the size of an hypothetical "cell" in the six-dimensional phase space and renders the integrations dimensionless. In non-relativistic dynamics Z_0 is both dimensionless and a scalar.[c] *By definition $F_0 = -kT \log Z_0$ is the "free energy"*, presumably extensive and carrying the units of energy.

An immediate application of (4.1) is to an *ideal* monatomic gas of N non-interacting spherical atoms of mass m in volume V. This could be any gas, provided it is sufficiently dilute that the effects of both interactions and of quantum statistics are negligible and can be ignored. An exact criterion is stated in later sections.

We now calculate Z_0 explicitly:

$$Z_0 = V^N (2\pi\hbar)^{-3N} \left(\int dp\, e^{-p^2/2mkT} \right)^{3N} = V^N \left(\frac{\sqrt{2\pi mkT}}{2\pi\hbar} \right)^{3N} . \qquad (4.6a)$$

The derived "free energy" F_0 has just one flaw: *it is not extensive.*

$$F_0 = -kTN \left[\log\left(V/v_0\right) + \frac{3}{2} \log\left(\frac{kT}{\varepsilon_0} \right) \right] . \qquad (4.6b)$$

To exhibit the various quantities in dimensionless form, we have multiplied and divided the ratio in (4.6a) by an arbitrary length L_0. The reference volume is $v_0 = L_0^3$ and the reference energy is $\varepsilon_0 \equiv \hbar^2 2\pi/mL_0^2$. By this artifice the argument of each of the two logarithms is rendered dimensionless while the *total* F_0 in (4.6b) remains rigorously (if not explicitly) independent of the actual value of L_0.[d] At constant density F_0 is dominated by $N \log N \gg N$.

---

[c]The present discussions do not extend either to special- or to general relativity. "Scalar" in the present context means invariant to the choice of coordinate system and all rotations, Galilean transformations, etc., thereof.

[d]The proof is left for the reader. For definiteness pick L_0 to be of the order of inter particle spacing, i.e. $L_0 = (V/N)^{1/3}$.

4.2.   Gibbs' Paradox and Its Remedy

If the free energy in statistical mechanics is to be a useful quantity and comparable to the thermodynamic free energy it *must*, among its other attributes, be extensive. F. W. Gibbs discovered the following paradox resulting from F_0, along with its resolution.

Suppose there are two identical containers of the identical gas at equal density, pressure and temperature. The total free energy must be $F_{2N} = 2F_0$, where F_0 is obtained from Z_0. Opening the partition between the two containers, allowing the $2N$ particles to occupy the joint volume $2V$, should not affect this result. But, as the reader will verify (using Eq. (4.6a or b)) the actual result is $F_{2N} = 2F_0 - 2NkT \log 2$. Given that the two methods of obtaining the free energy of the joint systems fail to agree, we can only conclude that the free energy *per* particle calculated using Eq. (4.6b) is ill-defined. To remedy this, Gibbs introduced his eponymous factorial into the definition of the *Canonical Partition Function*,

$$Z(N, V, T) = \frac{1}{N!} (2\pi\hbar)^{-3N} \prod_{\text{all } j} \iint d^3 q_j d^3 p_j \exp\{-\beta H\} . \qquad (4.7)$$

It is Z and not Z_0 which needs to be used. The resulting $F = -kT \log Z$ is *explicitly* extensive. In the case of the ideal gas (i.e. no interactions), the procedure of the preceding section yields:

$$F_{IG} = -kTN \left[\log\left(\frac{V}{Nv_0}\right) + \frac{3}{2} \log\left(\frac{kTe^{2/3}}{\varepsilon_0}\right) \right] , \qquad (4.8)$$

using Stirling's approximation (recall, $\log N! = N \log N - N$) and $e = 2.718\ldots$ If we use the free energy as given in (4.8) we now find the joint free energy of the two identical containers to be, correctly, twice the free energy of either one, whether or not they are connected.

Nevertheless it should be clear that if two *distinct* species were involved, according to the Second Law the mixture *should* have a lower joint free energy (and a higher total entropy) than did the separated gases, even though the total energy is unchanged when they are mixed. This is explored in Problem 4.1 and again in Chapter 7.

Problem 4.1. The joint partition function of two separate, non-interacting species is the product of the two: $Z = Z(N_A, V_A, T_A) \times Z(N_B, V_B, T_B)$. Calculate the total free energy when $T_A = T_B = T$. Calculate the "entropy of mixing" and the change in total energy after the two species are merged, using the total free energy derived from $Z(N_A, V_A + V_B, T) \times Z(N_B, V_A + V_B, T)$. What if the two species are indistinguishable? Discuss this in the context of the Second Law and the energy required in gaseous isotope separation.

The substantive difference between (4.6b) and (4.8) is that in the latter, the free energy per particle F_{IG}/N is a function of density ρ and temperature T only. (It is also possible to derive Eq. (4.8) from first principles using a different form for Z of a dilute gas, as shown in the next section.) The entropy is extracted in the usual way: $\mathcal{S} = -\partial F/\partial T$. Comparison of the calculations in (4.7) performed at given V and T with the formulas in (3.10) allow one to identify F with the quantity previously defined in Chapter 3 and in Eq. (4.2). Thus, the pressure in the ideal gas is,

$$p = - \left.\frac{\partial F}{\partial V}\right|_{N,T} = - \left.\frac{\partial F_{IG}}{\partial V}\right|_{N,T} = \frac{kTN}{V} \left(= kT\rho = \frac{kT}{v} \right) \qquad (4.9)$$

(precisely the ideal gas equation of state!) We also calculate $\mu = G_{IG}/N$ for the ideal gas. Replacing pV/N by kT,

$$\mu = \left.\frac{\partial F}{\partial N}\right|_{V,T} = kT + \frac{F_{IG}}{N} = \frac{3kT}{2} \log \left(\frac{2\pi\hbar^2}{mv^{2/3}kT} \right)$$
$$= -kT \left[\log \left(\frac{v}{v_0} \right) + \frac{3}{2} \log \left(\frac{kT}{\varepsilon_0} \right) \right], \quad \text{where } v = \frac{V}{N}. \qquad (4.10)^e$$

The condition for an ideal gas is an inequality: $\mu \ll -kT$. If at low T this inequality fails to be satisfied, either *quantum statistics* or *inter-particle interactions* cause the gas to condense into a liquid or superfluid or to transform into one of a variety of solids.

4.3.   The Gibbs Factor

The Gibbs factor should not be thought of as a correction, but rather as *mathematically required* in the normalization of the partition function of a classical gas of indistinguishable particles — as the following approach demonstrates. Allow a small neighborhood of each point in phase space $\alpha = (\mathbf{q}, \mathbf{p})$ to constitute a separate thermodynamical reservoir. In the absence of interactions the partition function is just a product of the Z_α's over the

---

[e]Strictly speaking, one should then use the equation of state to eliminate V/N in favor of kT/p in this expression to obtain $\mu(p, T)$.

distinct values of α. The total number of particles remains constrained at a given value, N. Using the by-now-familiar delta function:

$$Z = \frac{1}{2\pi} \int_{-\pi}^{\pi} dt\, e^{-itN} \prod_{\alpha} [1 + e^{(it-\beta \mathbf{p}^2/2m)} + \cdots]. \qquad (4.11a)$$

Here "..." refers to the sum of contributions $e^{2(it-\beta \mathbf{p}^2/2m)}$ from possible double occupation of $\mathbf{p}$, $e^{3(it-\beta \mathbf{p}^2/2m)}$ from triple occupation, etc. Regardless whether such multiple occupancies are permitted (in fact, they are prohibited for *fermions*), most momentum states are, in fact, *unoccupied* in a sufficiently dilute gas. Under the assumption that *no* momentum-state is occupied by more than one particle, one replaces (4.11a) by an expression that is identical to within neglected terms $O("...")$ and expands the exponential in powers of e^{it},

$$Z = \frac{1}{2\pi} \int_{-\pi}^{\pi} dt\, e^{-itN} \exp\left\{ \frac{V}{(2\pi\hbar)^3} \int d^3p\, e^{it-\beta \mathbf{p}^2/2m)} + \cdots \right\}$$

in which we replaced $\sum_{\alpha}$ by $\frac{V}{(2\bar{u}\hbar)^3} \int d^3p$,

$$= \frac{V^N}{(2\pi\hbar)^{3N} N!} \left[\int d^3p\, e^{-\beta \mathbf{p}^2/2m} \right]^N = Z_{IG}. \qquad (4.11b)$$

The logarithm of Z_{IG} now yields the correct F_{IG} as in Eq. (4.8).

Far from being an afterthought, Gibbs' factor ensures that each configuration of *indistinguishable* particles is counted just once. It provides evidence of the long shadow cast by the *correspondence principle*, according to which the quantum mechanics of identical particles devolves into classical theory in the limit $\hbar \to 0$.

4.4.   The Grand Ensemble

Fluctuations $O(\sqrt{N})$ in the number of particles N should make no observable difference in *any* observable result if N is sufficiently large. Moreover, allowing such fluctuations by fixing μ instead of N makes it that much easier to study *non*-ideal fluids. We start by verifying that the momentum states are not multiply-occupied, as claimed (without proof) in the preceding derivation.

The "trick" is simply to re-calculate (4.11b) by steepest descents. Let $it \equiv \tau$, a complex variable. We need to solve for the value of τ that renders the exponent stationary,

$$d/d\tau \left\{ -N\tau + \frac{V}{(2\pi\hbar)^3} \int d^3p \, e^{(\tau - \beta \mathbf{p}^2/2m)} \right\} = 0,$$

i.e.

$$N = V \int d^3p f^0(\varepsilon_p), \tag{4.12}$$

where $\bar{n}_p \propto f^0(\varepsilon_p) \equiv (2\pi\hbar)^{-3} e^\tau \exp{-(\mathbf{p}^2/2mkT)}$, is seen to be the famous Boltzmann distribution (cf. Eq. (4.14) below) once τ is identified as $\beta\mu$.

In evaluating Eq. (4.12) we find perfect agreement with the earlier Eq. (4.10), i.e. $\tau = \beta\mu = (3/2)\log(2\pi\hbar^2/mv^{2/3}kT)$. Because $\bar{n}_p \propto \exp\beta\mu$, $\beta\mu \ll -1 \Rightarrow \bar{n}_p \ll 1$ in a dilute gas. This, in turn, fully justifies the neglect of the terms indicated by "..." in (4.11a) provided $kT \gg 2\pi\hbar^2/mv^{2/3}$.

In studying the binomial expansion in Chapter 1 we found that a series could be approximated by its dominant term. The same holds for the partition function in Eq. (4.11) and motivates the replacement of Z by $\mathscr{Z}$, which is a simpler, unconstrained, product over α:

$$\mathscr{Z} = \prod_\alpha [1 + e^{(\beta\mu - \beta\mathbf{p}^2/2m)} + e^{2(\beta\mu - \beta\mathbf{p}^2/2m)} + \cdots]. \tag{4.13}$$

Except for trivial factors, this last is equivalent to $Z \exp(\beta\mu N)$. Note that each α includes all points in the volume V, i.e. an integral over space.

Let us write the free energy corresponding to the partition function in Eq. (4.13) as $\mathscr{F} = -kT \log \mathscr{Z} = F - \mu N$, differing from the original by a Legendre transformation. Where F was a function of $(N, V, T,)$ $\mathscr{F}$ is a function of (μ, V, T). Now the total number of particles hovers about a most probable value (N) determined by the choice of μ. This particular change of independent variables does not entail any substantial change in any of the physical results and for this reason one rarely distinguishes $\mathscr{F}$ from F. On the other hand, the new procedure *does* have one significant advantage.

Because $\mathscr{Z}$ is a product over distinct sectors, with no integration linking them, the evaluation of averages such as $\bar{n}_p$ allows treating each α independently of the others. Thus, when an average at $\mathbf{p}$ is performed, any term which does not refer to $\mathbf{p}$ cancels out. Explicitly:

$$\bar{n}_p = \frac{0 \cdot 1 + 1 \cdot e^{\beta\mu} e^{-\mathbf{p}^2/2mkT} + \cdots}{1 + e^{\beta\mu} e^{-\mathbf{p}^2/2mkT} + \cdots} \prod_{p' \neq p} \frac{1 + e^{\beta\mu} e^{-p'^2/2mkT} + \cdots}{1 + e^{\beta\mu} e^{-p'^2/2mkT} + \cdots}$$

$$\rightarrow \exp\beta\mu \exp{-\mathbf{p}^2/2mkT} + \cdots \tag{4.14}$$

This ensemble of states, in which all values of N are allowed, is called the *"Grand" canonical ensemble*, to distinguish it from the number-conserving *canonical ensemble* of the preceding sections. If instead of relaxing the conservation of N one required *greater* restrictions, the *microcanonical* ensemble discussed in the following problem would be of use. In the microcanonical ensemble, *both* N and E are strictly conserved and the calculations of thermodynamic properties from statistical mechanics are correspondingly more complicated.

Problem 4.2.   The *microcanonical* partition function W conserves energy and numbers of particles. Explicitly: $W = Tr\{\delta(E - H)\}$. Using the now-familiar representation of the delta function and the method of steepest descents valid at *large* N and E, show this $W \equiv \exp(\mathscr{S}(E)/k_B)$ at a stationary point $it = z$. Relate the "effective" temperature to E at the point of steepest descents by using $\frac{\partial \mathscr{S}(E)}{\partial z}\frac{\partial z}{\partial E} = \frac{\partial \mathscr{S}(E)}{\partial E}$ and invoking the definition of T in Eqs. (3.5) and (3.10). Now that E is "sharp", T is not. What is the distribution of T about its most probable value? of β?

4.5.   Non-Ideal Gas and the 2-Body Correlation Function

The most common two-body interactions involve the coordinates q_j but not the momenta. Therefore that part of the partition function that refers to the momenta of interacting particles is unchanged from its value in the ideal gas, and the effects of interactions are incorporated solely in the spatial configurations.[f] Specifically,

$$Z = Z_{IG} \cdot Q \cdot V^{-N}, \quad \text{where } Q \equiv \prod_{j=1}^{N} \int d^3 q_j \{e^{-\beta U(q_1, q_2, \ldots, q_N)}\} \qquad (4.15)$$

where $U(q_1, q_2, \ldots)$ is the total potential energy. Note: $QV^{-N} = \langle e^{-\beta U}\rangle_{vol}$.

However strong the interactions might be, by translational invariance the density of particles ρ is constant in the fluid phase — regardless whether it is liquid or vapor; this symmetry is broken only in the solid. Our proof starts

---

[f]According to the *ergodic hypothesis* (plausible but unproved), this procedure yields identically the same results as if we evaluated Z by first integrating the equations of motion starting from some set of initial conditions, and then averaged over initial conditions.

with the strict definition of ρ:

$$\rho(r) = \sum_{i=1}^{N} \langle \delta(r - r_i) \rangle_{TA} = \frac{N}{Q} \prod_{j=2}^{N} \int d^3 r_j \int d^3 r_1 \delta(r - r_1) e^{-\beta U(r_1, r_2, \ldots, r_n)}$$

in which the second equality follows from permutation symmetry. Because U is actually a function of just $N - 1$ vectors $r'_i \equiv r_{i1}$, $i = 2, \ldots, N$, we can take r_1 as the origin and integrate over r_1 getting V. Then Q is as follows:

$$Q = V \prod_{j=2}^{N} \int d^3 r'_j e^{-\beta U(r'_2, \ldots, r'_N)} . \qquad (4.15\text{a})$$

We transform the dummies of integration in ρ similarly, obtaining

$$\rho(r) = \frac{N}{Q} \prod_{j=2}^{N} \int d^3 r'_j e^{-\beta U(r'_2, \ldots, r'_n)} \int d^3 r_1 \delta(r - r_1) = \frac{N}{V} , \qquad (4.16)$$

a constant, *QED. Translational symmetry* is broken *only in solids.*

Whereas particle density is just a "one-body" property that distinguishes solids from fluids, the (dimensionless) 2-*body correlation* function $g(\mathbf{r})$ is more descriptive of correlations in the fluid phases. It is defined as,

$$g(r) = \frac{2V}{N(N-1)} \sum_{i}^{N} \sum_{j>i}^{N} \langle \delta(\mathbf{r} + \mathbf{r}_j - \mathbf{r}_i) \rangle_{TA} . \qquad (4.17)$$

In the fluid phase of interacting particles, $g(r)$ has to be isotropic. Hence g can only depend on the magnitude r of $\mathbf{r}$ and not on its orientation. If U vanishes (as it does in the ideal gas), the averaged quantity $\langle \delta(r - r_{ij}) \rangle_{TA} = 1/V$ is a constant (all intermolecular distances are equally likely), and $g_{IG} = 1$. Unless the forces are infinite-ranged g must approach this ideal-gas value asymptotically at large r. By permutation symmetry all terms in the above sum are equal after thermal averaging, therefore we need consider just one out of the $\frac{1}{2} N(N-1)$ distinct bonds. Then,

$$g(r) = V \langle \delta(\mathbf{r} - \mathbf{r}_2 + \mathbf{r}_1) \rangle_{TA}$$

$$= \frac{V}{Q} \prod_{j=3}^{N} \int d^3 r'_j \int d^3 r'_2 \delta(\mathbf{r} - \mathbf{r}'_2) e^{-\beta U(r'_2, r'_3, \ldots, r'_N)} \int d^3 r_1$$

$$= \frac{V^2}{Q} \prod_{j=3}^{N} \int d^3 r'_j \exp -\beta U(r, r'_3, \ldots, r'_N) . \qquad (4.18)$$

It is often convenient to consider, as we do now, only those simple cases where the total potential energy is the sum of two-body potentials and depend only on the intermolecular separations, i.e.

$$U = \frac{1}{2} \sum_i \sum_{j \neq i} u(r_{ij}) \,.$$

By an obvious symmetry,

$$\langle U \rangle_{TA} = \frac{N(N-1)}{2} \langle u(r_{12}) \rangle_{TA}$$

$$= \frac{N(N-1)}{2Q} \prod_{j=3}^{N} \int d^3 r u(r) \int d^3 r'_j \exp -\beta U(r, r'_3, \ldots) \int d^3 r_1$$

When combined with the 2nd line of (4.18) this yields,

$$\langle U \rangle_{TA} = \frac{N(N-1)}{2V} \int d^3 r u(r) g(r) \,, \tag{4.19}$$

an intuitively obvious result. Both $\langle U \rangle_{TA}$ and $g(r)$ are also functions of T. Including the kinetic energy, the total internal energy at constant N, V, and T is $E = \langle H \rangle_{TA} = (3/2)NkT + \langle U \rangle_{TA}$.[g] We shall try to calculate $g(r)$ shortly but — to motivate these arduous calculations — let us first relate $g(r)$ to the equation of state.

4.6.   The Virial Equation of State

As a straightforward application of the preceding one can derive an exact equation of state of the interacting fluid, expressing the pressure entirely as an integral over the two-body correlation function $g(r)$. Recall the definition in Eq. (3.10). Using (4.15),

$$p = kT \frac{\partial}{\partial V} \log \left(\frac{Z_{IG} Q}{V^N} \right) \Big|_T = kT\rho + kT \left(\frac{V^N}{Q} \right) \frac{\partial}{\partial V} \left(\frac{Q}{V^N} \right) . \tag{4.20}$$

The dependence of (Q/V^N) on V comes entirely from U, as each spatial integration is normalized by V.

---

[g] All calculations are explicitly for 3D. The changes required for d dimensions are obvious: $d^3 r \to d^d r$, $(3/2)NkT \to (d/2)NkT$ and $r_{ij} \propto V^{1/d} r_{0;i,j}$ in §4.6.

More specifically,

$$\frac{\partial}{\partial V} u(r_{ij}) = \frac{du(r_{ij})}{dr_{ij}} \frac{\partial}{\partial V} r_{ij} .$$

But

$$r_{ij} \propto V^{1/3} r_{0;i,j} ,$$

thus:[g]

$$\frac{\partial}{\partial V} r_{ij} = \frac{r_{ij}}{3V} .$$

By Eq. (4.15a), it follows that

$$\frac{\partial}{\partial V} \left(\frac{Q}{V^N} \right) = \frac{-N(N-1)}{6V^{N+1}kT} \int d^3 r'_2 \frac{du(r'_2)}{dr'_2} r'_2 \prod_{l=3}^{N} \int d^3 r'_l e^{-\beta U(r'_2, \dots)} \int d^3 r_1$$

$$= -\frac{N(N-1)Q}{6kTV^{N+2}} \int d^3 r \frac{du(r)}{dr} r g(r) . \tag{4.21}$$

This last is rigorous provided only that U is the sum of two-body radial potentials, $u(|\mathbf{r}_{ij}|)$. Subject to this mild restriction, the generic equation of state of any non-ideal fluid, vapor or liquid, becomes:

$$p = \rho kT \left[1 - \frac{\rho}{6kT} \int_0^\infty d^3 r \frac{du(r)}{dr} r g(r) \right] . \tag{4.22}$$

4.7. Weakly Non-Ideal Gas

Consider special cases: either of a dilute gas of molecules with arbitrarily strong, but short-ranged, interactions, or else of a fluid with weak, finite-ranged interactions at arbitrary density. In either case an expansion introduced by J. Mayer in the 1930's can be used to obtain leading correction terms to the ideal gas approximation.

Define: $\exp -\beta u(r_{ij}) \equiv 1 + f(r_{ij})$, where $f(r_{ij}) \equiv f_{ij} = \exp -\beta u(r_{ij}) - 1$ is a small, temperature-dependent quantity ranging from -1 at small r_{ij} (assuming a hard-core) to a small positive value in the region of inter-molecular attraction. Figure 4.1 on the next page compares the dimensionless quantities $-\beta u(r)$ and $f(r)$, at a relatively high temperature ($T = 2T_0$). For definiteness we have used the so-called "6–12" intermolecular potential in this example, a potential frequently used to fit the interactions of neutral,

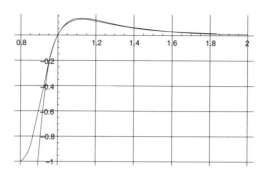

Fig. 4.1. f_{ij} **and "6–12" potential** $-u_{ij}/kT$, **versus** r/r_0 **in the range 0.8 to 2.**

The functions $f(x) = e^x - 1$ (upper curve) and $x \equiv -(1/t)(1/s)^6(1/s^6 - 1)$, are plotted versus $s = r/r_0$, where $t = T/T_0$ is the temperature in units of the potential and s is the intermolecular separation in units of the hard-core diameter $O(1\text{Å})$ (cf. Eq. (4.23)). The function $f(x)$ shown is calculated specifically for $t = 2$ but exhibits the features typical of high t: x and f are almost identical for $s > 1$ whereas, in the range $0 < s < 1$, $x \to -\infty$ and f remains finite and integrable, i.e. $f \to -1$.

spherically symmetric, atoms:

$$u(r) = kT_0 \left(\frac{r_0}{r}\right)^6 \left(\left(\frac{r_0}{r}\right)^6 - 1\right). \qquad (4.23)$$

This potential is attractive at distances greater than r_0, the hard-core diameter, and strongly repulsive at distances shorter than this. Its strength is given as kT_0, in temperature units. The related quantity $f(r)$ is close to $-\beta u(r)$ in the attractive region $r > r_0$ where $-u$ is positive, but unlike the latter (which diverges in $\lim \cdot r \to 0$), it saturates at -1 for $r < r_0$ and it is therefore integrable.

The calculation will make use of $I(T)$, the integral over $f(r)$:

$$I(T) = \int d^3 r f(r) \approx -\frac{4\pi}{3} r_0^3 \left[1 - \frac{2T_0}{3T}\right]. \qquad (4.24)$$

The approximate formula is a good stand-in for an exact evaluation. To derive it we retained just the leading terms in the high-temperature expansion, replacing f by $-\beta u(r)$ for $r > r_0$ and by -1 for $0 < r < r_0$. Although one can do better this type of simple formula is accurate enough for purposes of demonstration.

In addition to numerous complicated higher-order terms the configurational partition function Q incorporates a clearly identifiable leading power series in the "small" parameter f_{ij}. Knowing it must be an exponential form, we try to guess what it is:

$$\frac{Q}{V^N} \equiv V^{-N} \prod_{i=1}^{N} \prod_{j \neq i}^{N} \int d^3 r_i [1 + f_{ij}]$$

$$= V^{-N} \prod_{i=1}^{N} \int d^3 r_i \left[1 + \sum_{i,j} f_{ij} + \frac{1}{2!} \sum_{(i,j)} f_{ij} \sum_{(k,l) \neq (i,j)} f_{kl} + \cdots \right] \quad (4.25)$$

This expansion leads itself to a diagrammatic expansion. If this is to your taste, see the books in footnote h. Evaluation of the first few terms shows the leading contribution to be,

$$\frac{Q}{V^N} \approx \exp \left[\frac{N(N-1)}{2V} I(T) \right] ; \quad \text{hence } p = \rho k T \left[1 - \frac{1}{2} \rho I(T) \right]. \quad (4.26)$$

The *virial* expansion of p is defined as a power series in the small quantity ρ, with T-dependent coefficients (the so-called *virials*):

$$p = kT[\rho + \rho^2 B(T) + \rho^3 C(T) + \cdots] \quad (4.27)$$

Comparing the two preceding equations we deduce that the *second virial coefficient* $B(T) = -I(T)/2$, with $I(T)$ defined (and then approximated) in Eq. (4.24). Calculation of the *third* virial coefficient $C(T)$ requires integrations over three-body clusters such as $f_{ij} f_{jk}$, etc. However with just the aid of (4.26) we are already able to compare these results with van der Waals' intuitive equation of state, Eq. (3.17). Inserting (4.24) for $I(T)$ into (4.26):

$$p = \frac{kT}{v} \left(1 + \frac{b}{v} \right) - a \left(\frac{1}{v} \right)^2, \quad \text{with } b = \frac{v_0}{2} \quad \text{and } a = \frac{v_0 kT_0}{3}. \quad (4.28)$$

Here we defined $v_0 \equiv 4\pi r_0^3/3$, the effective atomic volume. Aside from any additional corrections $O(v^{-3})$ due to higher terms in the virial expansion, Eq. (4.28) *agrees formally* with Eq. (3.17) if in the latter we allow van der Waals' b to depend on v as follows: $b = v_0/(2 + v_0/v)$. This may serve to explain why b is *not* independent of v and T in real gases (an experimental fact that was noted by van der Waals with some chagrin).

Explicit but formal expressions, in terms of integrals over "irreducible clusters" of m particles, have been obtained for all virial coefficients.[h] But two factors: the technical difficulty of computing many-body cluster integrals

[h]D. H. Goodstein, *op. cit.*, Chapter 4; J. Mayer and M. Mayer, *Statistical Mechanics*, J. Wiley, New York, 1940; H. L. Frisch and J. L. Lebowitz, *The Equilibrium Theory of Classical Fluids*, Benjamin, New York, 1964.

and the knowledge that, *even if all the coefficients were known*, the virial expansion converges only up to the first singularity — whether it occurs at the critical point, the triple point, or somewhere in the complex ρ plane,[i] have stymied widespread applications. Without a valid closed-form expression p cannot be analytically continued into the condensed regions.

4.8. Two-body Correlations

We return to $g(r)$. When the structure of materials is probed by elastic X-Ray or elastic neutron diffraction, it is the built-in correlations that are observed. After being scattered just once by the entire material, the diagnostic particle — whether it is an X-Ray, a neutron, or an electron — has momentum $\hbar\mathbf{q}$ imparted to it. The phase coherence of the scattering of the incident beam $\mathbf{k}$ to $\mathbf{k} + \mathbf{q}$ is proportional to $I(\mathbf{q})$. In the lowest-order Born approximation,

$$I(\mathbf{q}) = \left\langle \left| \sum_{j=1}^{N} e^{iq \cdot r_j} \right|^2 \right\rangle_{TA} . \tag{4.29a}$$

This expression can also be written in more transparent form: $I(\mathbf{q}) = \int d^3 r e^{i\mathbf{q}\cdot\mathbf{r}} \phi(r)$, this being the Fourier transform of $\phi(\mathbf{r}) = N\delta(\mathbf{r}) + \sum_i^N \sum_{j\neq i}^N \langle \delta(\mathbf{r} + \mathbf{r}_{ij}) \rangle_{TA} \equiv N\delta(\mathbf{r}) + \frac{N(N-1)}{V} g(r)$ with $g(r)$ the two-body correlation function as defined in (4.17). So,

$$I(\mathbf{q}) = N \left[1 + \rho \int d^3 r e^{i\mathbf{q}\cdot\mathbf{r}} g(r) \right] . \tag{4.29b}$$

After excluding forward scattering ($q = 0$) this expression simplifies further in terms of $h(r) = g(r) - 1$. The *structure factor* $S(q)$ is defined as,

$$S(\mathbf{q}) = \frac{I(\mathbf{q})}{N} = \left[1 + \rho \int d^3 r e^{i\mathbf{q}\cdot\mathbf{r}} h(r) \right] , \quad \text{for } \mathbf{q} \neq 0 . \tag{4.30}$$

Because $g \to 1$ asymptotically, $h(r)$ vanishes at large r.

$S(\mathbf{q})$ mirrors $g(r)$ in reciprocal space. It is isotropic in fluids and in an ideal gas, $S(q) = 1$. In denser fluids $S(q)$ is computed from (4.30) using a known $h(r)$, or it is obtained directly from experiment. Measurements on a

---

[i]J. Groeneveld, *Phys. Lett.* **3**, 50 (1962). But for a *practical* examination of the properties of fluids see J. Stephenson, *Phys. Chem. Liq.* **10**, 229 (1981) and "Hard- and Soft- Core Equations of State", in *Proc. 8th Symposium on Thermophysical Properties*, Vol. 1, J. Sengers, Ed., ASM, New York [book # 100151], pp. 38–44.

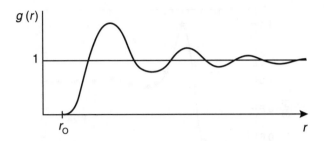

Fig. 4.2.   Two-body correlation $g(r)$ for a highly correlated liquid.
(Schematic plot). $g(r)$ vanishes at separations less than the hard-core diameter and
first peaks where the attractive well is deepest. The asymptotic value at $g(r) = 1$ is
also the two-body correlation function of a very dilute (i.e. "ideal") gas at high T.

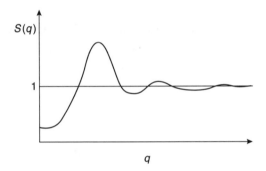

Fig. 4.3.   Structure factor $S(q)$ of a normal fluid.
(Schematic plot). At finite $T \lim \cdot q \to 0$, $S(q) \neq 0$ (see text).
At large q, $S(q) \to 1$.

few widely different fluid systems are displayed below, either schematically
or accurately.

The schematic plot of $g(r)$ in Fig. 4.2 indicates that this spherically symmetric function vanishes for $r < r_0$, peaks at the bottom of the attractive
well at r_m and exhibits smaller peaks at higher-"shell" diameters. In a weakly
interacting or dilute gas only the first peak would be observed.

The Structure Factor, a.k.a. the Fourier transform $S(q)$ of $g(r)$ defined
in Eq. (4.29b) is shown (also schematically) in Fig. 4.3.

It can be proved that in $\lim \cdot q \to 0$, $S(q) \to \rho k T \kappa_T$; thus this
plot contains valuable thermodynamic as well as structural information.
Figure 4.4 illustrates similar behavior in a superfluid (^{4}He) just below the
critical temperature for the onset of superfluidity.

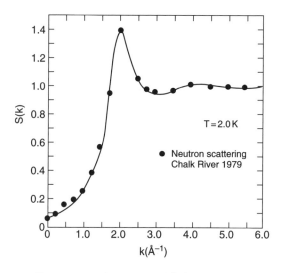

Fig. 4.4. Structure function $S(k)$ for superfluid ^{4}He.

Note 1: $S(k) \to 1$ at large k for a superfluid also.

Note 2: $S(0) \neq 0$ at finite T.

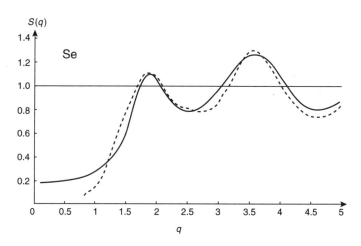

Fig. 4.5. Structure function of Amorphous Selenium.[j]

(experimental $-----$, modeled $\overline{}$)

Exceptionally some solids, such as amorphous selenium (Se), display liquid-like features. Figure 4.5 suggests graphically that at least some glasses are, in essence, "frozen liquids".

[j]Adapted from S. C. Moss and D. L. Price, 1985.

All these examples indicate that gases, fluids, superfluids and glasses all have similar, isotropic, diffraction patterns. (The diffraction patterns of crystalline materials differ sharply from these, as we shall soon see). The question at hand is, how do we calculate or estimate g or h in a fluid, to understand the behavior in somewhat greater detail?

In the dilute gas we might guess $g(r) \equiv g_0(r) \approx \exp -\beta u(r)$ as in the "barometer equation", normalized to satisfy the boundary conditions $g_0 = 0$ at $r = 0$ and $g_0 \to 1$ asymptotically. Insertion into the virial equation of state Eq. (4.22) shows this *does* yield the identical second virial coefficient $B(T)$ as that found in Eqs. (4.26) and (4.27):

$$-\frac{1}{6kT} \int d^3r \frac{du(r)}{dr} r e^{-\beta u(r)} = \frac{1}{6} \int d^3r \, r \frac{d}{dr}(e^{-\beta u(r)} - 1);$$

after a partial integration this is,

$$-2\pi \int_0^\infty dr \, r^2 (e^{-\beta u(r)} - 1) = -\frac{1}{2} \int d^3r f(r) = -I(T)/2.$$

This checks out for a very dilute gas. But even though $\exp -\beta u(r)$ vanishes for $r < r_0$ and peaks at r_m, this approximation to $g(r)$ lacks the subsequent maxima that are indicative of long-range multiparticle correlations.

It has been shown rigorously that the systematic expansion of $g(r)$ in powers of the density can be cast in the form,[h,i]

$$\log g(r) + \frac{u(r)}{kT} = \sum_1^\infty \delta_m(r)\rho^m, \tag{4.31}$$

with $\delta_m \equiv m$-fold integral over clusters of $m+1$ f's (not to be confused with the Dirac function.) The leading order correction is,

$$\delta_1(r_{12}) = \int d^3r_3 f_{13} f_{23}. \tag{4.32}$$

Problem 4.3. *Evaluation of δ_1.* Generalize Eq. (4.24) to $I_k(T) = \int d^3r e^{i k \cdot r} f(r)$, calculating $I_k(T)$ to the same accuracy as $I(T)$ in the text. Invert the Fourier expansion to obtain f in terms of I_k, then express $\delta_1(r)$, defined above, as an integral over $|I_k|^2$ and determine the positions of its maxima relative to r_m. (They reflect the maxima in g. Are they in the "right" place?) Use this result to estimate the *third* virial coefficient, $C(T)$ in Eq. (4.27).

Problem 4.4. Obtain $I(T)$, $I_k(T)$, $B(T)$ and $C(T)$ for *hard* (impenetrable) *spheres* of radius r_0. Compare your results with the known[k] values: $B(T) = 2\pi r_0^3/3$, $C(T) = (5/8)(B(T))^2$.

In the next order,

$$\delta_2(2_{12}) = \frac{1}{2!} \iint d^3 r_3 d^3 r_4 f_{13} f_{34} f_{24} (2 + f_{23} + f_{14} f_{23}) . \tag{4.33}$$

The two corrections (4.32) and (4.33) allow to calculate and verify the second- and third-shell peaks in g and in S of highly correlated fluids and to evaluate the next higher virial coefficient. Higher-order corrections δ_m can *all* be obtained geometrically from "irreducible clusters", in a tour de force diagrammatic representation of the series expansion of Eqs. (4.25) and (4.31).[h,k] We do not dwell on this because, near critical points, no finite number of correction terms ("diagrams") in the power expansion will suffice. Instead, two distinct schemes have evolved over time, each seeking to sum the "most relevant" terms in this series up to infinite order: the hypernetted chain and the Percus–Yevick equations.[l] The advantage of either of these approximations is that the *form* of the solution allows analytic continuation of the themodynamic functions through the vapor-liquid phase transition and beyond.

Even though the magnitude of the errors which result from omission of "less relevant" terms cannot be quantitatively assessed, the hypernetted chain and the Percus–Yevick equations marked an important development in physical chemistry. They encouraged the theoretical study of multi-component fluids and electrolytes.[h] Further generalizations have allowed the study of inhomogeneities in liquid solutions caused by surfaces and, most importantly, of the "Helmholtz layer" which forms in electrolytes near electrodes. But, given the uncertainties and difficulties in an analytical approach, contemporary trends have led elsewhere: to molecular dynamics and to "Monte Carlo" simulations, in which computer "experiments"

---

[k]L. Boltzmann (1899): see H. Happel, *Ann. Physik* **21**, 342 (1906). Beyond these, the next few virial coefficients for hard spheres have been known for some decades; B. Nijboer and L. van Hove, *Phys. Rev.* **85**, 777 (1952) and for the fifth and sixth, F. H. Ree and W. G. Hoover, *J. Chem. Phys.* **40**, 939 (1964).

[l]The original derivations and the solutions of these two equations — which, unfortunately, are not in perfect agreement — are reprinted, together with variants and useful commentary, in the compendium by Frisch and Lebowitz.[h] The most up-to-date explanations are found in N.H. March and M. P. Tosi, *Introduction to Liquid State Physics*, World Scientific, 2002.

simulate physical reality and output the required processed data without uncontrolled error.

Two exceptions stand out: one-dimensional fluids, for which we calculate the partition function in closed form, and three-dimensional solids in the low-temperature (harmonic) regime. However, in the latter the statistical mechanics of phonons and of vacancies, interstitials, dislocations and other structural defects requires quantum mechanics. So first let us see what makes 1D systems distinctive.

4.9.   Configurational Partition Function in 1D[m]

In 1936, L. Tonks first calculated the configurational partition function of a gas "of elastic spheres" in 1D, i.e. the one-dimensional gas with hard-core repulsions. Here the Boltzmann factor $\exp -\beta U$ is trivial: it is $\equiv 0$ whenever *any* two particles come too close, i.e. whenever any pair $|x_j - x_i| < b$, and it is $= 1$ otherwise. The ordering of the particles $\cdots < x_n < x_{n+1} < \cdots$ can never change as they cannot "get around" one another. Hence,

$$Q = N! \int_{(N-1)b}^{L} dx_N \int_{(N-2)b}^{x_N - b} dx_{N-1} \cdots \int_{b}^{x_3 - b} dx_2 \int_{0}^{x_2 - b} dx_1 1, \qquad (4.34a)$$

upon picking any representative order. Next, a change of variables simplifies the integration: let $y_n = x_n - (n-1)b$. Then,

$$Q = N! \int_{0}^{L-(N-1)b} dy_N \int_{0}^{y_N} dy_{N-1} \cdots \int_{0}^{y_3} dy_2 \int_{0}^{y_2} dy_1 1 = (L - (N-1)b)^N$$

$$= L^N (1 - \rho b)^N, \qquad (4.34b)$$

with $\rho = N/L$. Therefore the free energy of Tonks' gas is precisely $F_{\text{Tonks}} = F_{IG} - NkT \log(1 - \rho b)$. Upon differentiating *w.r.* to L at constant N and T one obtains Tonks' equation of state,

$$p = \rho kT + \rho^2 bkT/(1 - \rho b) = \rho kT/(1 - \rho b). \qquad (4.35)$$

Adding a *constant, infinite-ranged*, two-body attractive interaction $u_{ij} = -2a(N-1)/L$ changes $F_{\text{Tonks}} \Rightarrow F_{vdw} = F_{\text{Tonks}} - aN(N-1)/L$. Differentiation of F_{vdw} *w.r.* to L *yields van der Waals' equation of state*, Eq. (3.17), *precisely!* (A similar approach in 3D yields a similar result). But the very artificiality of this *force-free*, infinite-ranged two-body potential u_{ij} shows

---

[m]References and calculations of some exact correlation functions in 1D are given in D. Mattis, *The Many-Body Problem*, World Scientific, 1994, Chap. 1.

van der Waals' equation of state to have been the expression of a mean-field "toy model" — rather than the law of Nature he so dilligently sought.

Takahashi was first to recognize that one could include a more physically sensible *short-ranged* potential and *still* be able to evaluate Q. Let it vanish at separations exceeding $2b$, so that only nearest-neighbors atoms interact, as before. Q then takes the form:

$$Q = N! \int_0^{L^*} dy_N \cdots \int_0^{y_3} dy_2 \int_0^{y_2} dy_1 \exp -\beta \sum_{n=1}^{N-1} u(y_{n+1} - y_n) \qquad (4.36)$$

where $L^* \equiv L - (N-1)b$ is the "effective" length of the chain. This has the form of an iterated convolution, the type of multiple integration best evaluated by Laplace transform. Let $Q/N! \equiv C(L^*)$. The Laplace transform of $C(L^*)$ is,

$$\bar{C}(s) = \int_0^\infty dL^* C(L^*) e^{-sL^*} = s^{-2} K(s)^{N-1} \qquad (4.37)$$

where $K(s) = \int_0^\infty dy\, e^{-sy - \beta u(y)}$ and s is determined *via* the equation, $-\frac{\partial}{\partial s} \log K(s) = \langle y \rangle = L^*/N$; the *equation of state* is constructed by combining this equation with the usual definition of p. The inverse Laplace transform of the *rhs* of (4.37) shows Q to be analytic in β; even if there is a phase transition it occurs at $T = 0$ ($\beta = \infty$) and the system is thus in its high-T phase at all finite T.

The multiple integral (4.36) can also be evaluated by defining a *transfer matrix* M. We examine this in some detail, for similar procedures prove useful in later applications. Assume that after $n-1$ integrations the result is proportional to a normalized $\Psi_n(y_n)$; then, $\int_0^{y_{n+1}} dy_n \exp -\beta u(y_{n+1} - y_n) \Psi_n(y_n) = q_n \Psi_{n+1}(y_{n+1})$. Here, using a step function ($\theta(x) = 0$ for $x \leq 0$, $\theta(x) = 1$ for $x > 0$,) the *transfer matrix* is: $M(y_{n+1} - y_n) \equiv \theta(y_{n+1} - y_n) \exp -\beta u(y_{n+1} - y_n)$. By iteration,

$$Q = N! \prod_{n=1}^{N-1} q_n \,.$$

This integration turns into an eigenvalue problem once we posit M has a complete set of eigenfunctions Φ in which the Ψ's can be expanded. As the integral is iterated, the coefficient of the eigenfunction Φ_0 with the largest eigenvalue q_0 grows the fastest and ultimately dominates. Hence after $n \gg 1$ iterations, $q_n \to q_0$ and Ψ_n tends to Φ_0. This leads to the eigenvalue equation:

$$\int_0^\infty dy' M(y - y') \Phi_0(y') = q_0 \Phi_0(y) \qquad (4.38)$$

for the largest $q \equiv q_0$. Once found, we can set $\log Q = \log N! + N \log q_0$ (+ inessential terms lower-order in N,) and the problem is solved; indeed, the value of q_0 is deduced below.

However the eigenfunction $\Phi_0(y)$ is itself of some interest. It is non-negative (the integrands in (4.36) are all non-negative) and once normalized, such that $\int_0^\infty dy \Phi_0(y) = 1$, it is identified as the probability distribution for the nearest-neighbor inter-particle separation y. For $\Phi(y)$ to be normalizable it must vanish sufficiently fast at large y to allow the upper limit of the integrations to be set at ∞. Also, $\Phi(0) = 0$. (This follows immediately from Eq. (4.38) upon noting that M vanishes identically wherever its argument is negative).

Define

$$F(s) = \int_0^\infty dy \Phi_0(y) \exp -(sy)$$

and insert into Eq. (4.38):

$$\int_0^\infty dy \int_0^\infty dy' [M(y - y')e^{-s(y-y')}]e^{-sy'} \Phi_0(y') = q_0 F(s) .$$

The integration over y on the *lhs* can be shifted into an integral over $z \equiv y - y'$. Mindful that $M(z)$ vanishes for $z < 0$ one finds $F(s)K(s) = q_0 F(s)$. Hence $q_0 = K(s)$, recovering the earlier results in $\lim \cdot N \to \infty$.

4.10.   One Dimension versus Two

Whereas in 1D there are several complementary ways to understand the lack of any long-range order (LRO), the proofs in 2D are both subtler and less universal.

Consider the famous proof, attributed to Landau, of the absence of LRO in 1D. If one severs an arbitrary linear array assumed to possess LRO by breaking it into two uncorrelated segments, the energy cost Δe is independent of the size of the system. However, assuming there are N equivalent places where the cut can take place, the entropy is enhanced by $k \log N$ by a single cut. It follows that, at any finite T, the change in free energy $\approx \Delta e - kT \log N$ is always favorable to this break-up.

Subsequently each severed part can be cut in turn, each cut again lowering the over-all free energy. This continues until the average, individual, uncorrelated segments are no bigger than $e^{\Delta e/kT}$, at which point the free energy is at a minimum. Therefore at finite T there is no LRO in 1D and even short-range order (SRO) typically decays exponentially with distance. When

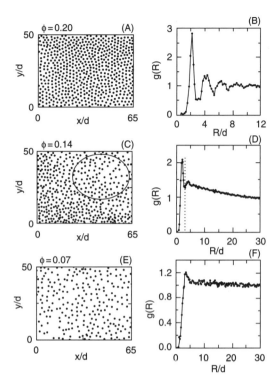

Fig. 4.6.   Liquid-vapor phase transition in 2D.[n]

(A), (C) and (E) are snapshots of the positions of millimeter-sized like-charged metal balls at three densities $\phi = \rho d^2$. (B), (D) and (F) are plots of the two-body correlation functions $g(R)$ at the same three densities. (The dashed line in (D) indicates the average inter-particle distance at $\phi = 0.14$.)

combined with the preceding section, where we showed that the largest eigenvalue of the transfer matrix to be analytic in β, this is convincing evidence that *even if* an ordered phase existed at $T \equiv 0$, in thermal equilibrium *only* the high-temperature phase exists in any 1D system at finite T. In 2D the simple version of Landau's argument is invalidated, as the energy required to destroy LRO by cutting the plane rises to $O(\Delta e \sqrt{N})$ while the entropy gain remains at $O(k \log N)$.

Examination of long-wavelength excitations shows the actual situation to be more nuanced. If the dynamical variables (such as the distance between nearest-neighbor (*n-n*) atoms or the angle between *n-n* spins' orientation) are continuous there can be no LRO at any finite T. If, however, there is an

[n] Adapted from B. V. R. Tata *et al.*, *Phys. Rev. Lett.* **84**, 3626 (2000).

energy gap against elementary excitations (as in the Ising model or in the lattice gas), the ground state LRO *may* persist to finite T and one may need to account for several distinct phases, including some with LRO.

But even the absence of LRO does not necessarily preclude phase transitions. In 2D, two distinct fluid phases may be allowed — even if the forces are purely repulsive. A striking demonstration is shown in Fig. 4.6 for like-charged metal balls, each of diameter $d = 1.59$ mm, on a 2D "table". The two-body correlation function changes character from correlated liquid to dilute gas as a dimensionless density $\phi = \rho d^2$ is lowered from 0.2 to 0 at fixed T.

4.11. Two Dimensions versus Three: The Debye–Waller Factors

It is fortunate for us that the world as we know it is imbedded in *at least* 3 spatial dimensions (and perhaps as many as 10 or 11 according to some theorists), as the very existence of elastic solids in *fewer* than 3 dimensions is problematical. In fact, even if one assumed LRO in a given lattice at low temperatures in $d < 3$ dimensions, either quantum or thermal lattice vibrations destroys it. The importance of fluctuations in low dimensions can already be understood within the context of the structure factor. Let us start in 3D.

Recall the definition of $S(\mathbf{q})$ in Eqs. (4.29) and (4.30):

$$S(\mathbf{q}) = \frac{1}{N} \left\langle \left| \sum_{j=1}^{N} e^{i q \cdot r_j} \right|^2 \right\rangle_{TA}.$$

Now, in an "ideal" 3D solid the positions r_j of individual atoms are $R^0(n_1, n_2, n_3) = n_1 t_1 + n_2 t_2 + n_3 t_3$, with the $\mathbf{t}_\alpha$ a set of non-coplanar "primitive translation vectors" and the n_α's arbitrary integers. Starting from any cell one can find the position of any other cell, however distant, by an appropriate choice of three n_α's; that is the nature of a space lattice with LRO. It is then simple to calculate $S_0(\mathbf{q})$, the *ideal* structure factor.

First, one defines three non-coplanar primitive translation vectors of the "reciprocal lattice",

$$K_1 = 2\pi \frac{t_2 \times t_3}{t_1 \cdot t_2 \times t_3}, \quad K_2 = 2\pi \frac{t_3 \times t_1}{t_1 \cdot t_2 \times t_3}, \quad K_3 = 2\pi \frac{t_1 \times t_2}{t_1 \cdot t_2 \times t_3}. \quad (4.39)$$

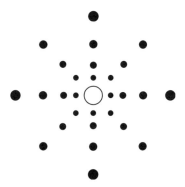

Fig. 4.7.   Laue back-reflection pattern of oriented AgCl crystal.
X-Ray pattern of a single crystal silver chloride oriented along a principal (100)
direction.

They have the property that any point in the reciprocal lattice can be written in the form $\mathbf{q}_0(m_1, m_2, m_3) = m_1 K_1 + m_2 K_2 + m_3 K_3$, with m_α's arbitrary integers. For *any* and all such q_0's, terms of the form $\exp(i q_0 \cdot \mathbf{R}^0) = \exp(i 2\pi \times \text{integer}) = 1$. That makes $S_0(\mathbf{q}) = N$ if $\mathbf{q} = \mathbf{q}_0$ and $S_0(\mathbf{q}) = 0$ if $\mathbf{q} = \mathbf{q}_0$. Indeed, the X-Ray diffraction pattern of an ideal solid does consist of discrete Laue spots, each characterized by a different value of $\mathbf{q}_0$. Once a few have been tallied it is easy to extract the three primitive vectors K_α and by inversion of (4.39), to obtain the structure and geometry of the space lattice.

This discrete spectrum $S_0(\mathbf{q})$, of which a typical example is shown in Fig. 4.7, differs noticeably from the isotropic, continuous, structure factor in fluids shown in Figs. 4.3–4.5. The discreteness is a direct consequence of LRO. Now, what if the actual atomic positions differ from their ideal values? We shall see that the answer depends crucially on the number of dimensions. But first, the usual theory in 3D as it is recounted in textbooks on solid-state physics.

At finite T the atomic positions are shifted from their ideal values, i.e. $r_j = R_j^0 + \delta r_j$. The structure factor becomes,

$$S(\mathbf{q}) = 1 + \frac{1}{N} \left\langle \sum_{j=1}^N e^{iq \cdot r_j} \sum_{\substack{j'=1 \\ j' \neq j}}^N e^{-iq \cdot r_{j'}} \right\rangle_{TA}$$

$$= 1 + \sum_{j \neq 0}^N e^{iq \cdot R_j^0} \left\langle e^{iq \cdot (\delta r_j - \delta r_0)} \right\rangle_{TA}, \tag{4.40}$$

setting j' as the origin ($n_1 = n_2 = n_3 = 0$). Expand δr in normal modes,

$$\delta r_j = \frac{1}{\sqrt{N}} \sum_k e^{i\mathbf{k}\cdot\mathbf{R}_j^0} u(\mathbf{k}), \quad \text{and } \delta r_j - \delta r_0 = \frac{1}{\sqrt{N}} \sum_k (e^{i\mathbf{k}\cdot\mathbf{R}_j^0} - 1) u(\mathbf{k}). \quad (4.41)$$

Only those normal modes with excursion $\mathbf{u}$ parallel to $\mathbf{q}$ contribute to the exponent. If the forces are perfectly elastic, as one may assume for small excursions, thermal averaging expresses the potential energy of each normal mode as a quadratic form. This is inserted in the Boltzmann exponent, $\exp -(\beta/2) M k^2 s^2 |u(k)|^2$, where M is the mass of the unit cell and s the speed of sound. Thus,

$$\langle e^{iq\cdot(\delta r_j - \delta r_0)} \rangle_{TA} = \frac{\prod_k \int du(k) e^{-\frac{1}{2}\beta M s^2 k^2 |u(k)|^2} e^{iq\cdot(\delta r_j - \delta r_0)}}{\prod_k \int du(k) e^{-\frac{1}{2}\beta M s^2 k^2 |u(k)|^2}}. \quad (4.42)$$

Substituting (4.41) in the exponent and completing the square yields us the *Debye-Waller* factor for *elastic scattering* of X-Rays,

$$\langle e^{iq\cdot(\delta r_j - \delta r_0)} \rangle_{TA} = \exp - \left(\frac{q^2 kT}{M s^2} \frac{1}{N} \sum_{k \subset BZ} \frac{(1 - \cos \mathbf{k}\cdot\mathbf{R}_j^0)}{k^2} \right). \quad (4.43a)$$

The sum over k is constrained to the first *Brillouin Zone*, as the first unit cell about the origin in reciprocal space is called. Its volume $K_1 \cdot K_2 \times K_3$ contains precisely N points k, reflecting the number of cells in the space lattice. The sum in the exponent is named after Ewald, a pioneering practitioner of X-Rays, and can be identified as the Coulomb potential of discrete charges placed on an equivalent lattice. For present purposes it is unnecessary to know it in detail; we are just investigating its singularities. In the thermodynamic limit it can be approximated by a simpler integral,

$$\langle e^{iq\cdot(\delta r_j - \delta r_0)} \rangle_{TA} = \exp - \left(\frac{(qa_0)^2 kT a_0}{M s^2 (2\pi)^3} \int_{BZ} d^3 k \frac{(1 - \cos \mathbf{k}\cdot\mathbf{R}_j^0)}{k^2} \right) \quad (4.43b)$$

where $a_0^3 = t_1 \cdot t_2 \times t_3$ is the volume of a unit (space) cell. At relatively small distances R_j^0 the integral is bounded but *does* grow as R^2.

Asymptotically, at large R, it is permissible to neglect the oscillatory $\cos \mathbf{k}\cdot R$ term altogether and the integral reaches its ultimate value, which is:

$$\exp - \left(\frac{(qa_0)^2 kT}{M s^2} \frac{a_0}{(2\pi)^3} \int_{BZ} d^3 k \frac{1}{k^2} \right) \approx \exp - \left(\frac{(qa_0)^2 T}{\theta_D} \right), \quad (4.44)$$

independent of R. Here we have introduced the "Debye theta" θ_D, a quantity having the units of temperature. Multiplied by Boltzmann's constant,

$k\theta_D \approx O(Ms^2)$ it marks the boundary separating classical from quantum behavior in the dynamics of a solid.[o] θ_D is listed for numerous materials in specialized handbooks and is known to range from a few degrees Kelvin in the softest material to a few hundred or 1000's in the hardest. Classical statistical mechanics, used in Eqs. (4.42) and (4.43) to evaluate the thermal average, is valid in the high-temperature regime $T > \theta_D$.

Although the sum over j in Eq. (4.40) still restricts the diffraction to those discrete values of $\mathbf{q}$ which coincide with the reciprocal lattice, once *inelastic* scattering of the X-Rays is included the calculated Laue spots acquire a finite breadth — also observed in experiments.[P] Summarizing these results: with increasing temperature the low-order points become broader while high-order (large q_0) Laue spots become so broad that they virtually disappear. The X-Ray picture Fig. 4.7 illustrates the increased breadth of the Laue points with distance from the origin.

Now we turn to 2D. Modifying (4.44) for the dimensionality, one has

$$\exp - \left(\frac{(qa_0)^2 kT}{Ms^2} \frac{1}{(2\pi)^2} \int_{BZ} d^2k \frac{1}{k^2} \right) = 0, \tag{4.45}$$

vanishing because of an infrared divergence (i.e. at small k).[q] Therefore the summand in Eq. (4.40) vanishes in the asymptotic region. This limits the number of terms which contribute to the sum in Eq. (4.40) to the sites closest to the origin.

To estimate $S(q)$ in this case let us take an extreme view and assume that the thermal averaged factor takes the value $\exp -[\alpha(qa_0)^2]$ at n.-n. sites and effectively vanishes beyond the first shell. Then evaluating Eq. (4.40) for the square (sq) lattice one obtains the structure factor

$$S(q) = 1 + 2(\cos q_x a_0 + \cos q_y a_0)e^{-\alpha(qa_0)^2}$$

$$\approx 1 + 2 \left(2 - \frac{1}{2}(qa_0)^2 \right) e^{-\alpha(qa_0)^2}. \tag{4.46}$$

This function is continuous, virtually isotropic, and tends to 1 at large q. Although derived for an ordered state, the resulting $S(q)$ is, for $qa_0 = 1$, closer to that of a fluid than to the discrete spectrum of a crystal.

---

[o] Again we use k_B without the subscript, but it should not be confused with the momentum dummy variable k in the integrations.
[P] The reason: instead of $\mathbf{q}_0$ one may have $\mathbf{q}_0 \pm \delta\mathbf{q}$ owing to the emission or absorption of a phonon of wavevector $\delta\mathbf{q}$, with $\delta\mathbf{q}$ ranging over the BZ.
[q] In 1D the IR divergence is even more severe!

This result is not a fluke. In a later chapter we shall confirm the lack of LRO at finite T in *all* systems having a continuous symmetry (such as phonons) in sufficiently low-dimensions.

Chapter 5

The World of Bosons

5.1.   Two Types of Bosons and Their Operators

Nature seems to favor particularly two types of particles: fermions and bosons. Other types are few and far between. This chapter deals only with bosons. Oscillations of coherent systems, bearing such names as "phonons" in ordered materials, "photons" in the electromagnetic field, "magnons" in magnetic systems, "plasmons" in charged fluids, "gluons" in high-energy physics, etc., are all typically bosons. However these constitute just one type of boson.

Even numbers $(2, 4, \ldots)$ of fermions bound together with a large binding energy, such as alpha particles and helium or argon atoms, constitute a different genus of boson. Whereas oscillations can be freely created or destroyed, the number N of atoms is conserved — at least in the usual contexts where thermodynamics is applicable. Number conservation can be an issue unless it is amicably resolved, e.g. by the introduction of a chemical potential. Such reconciliation cuts a little deeper in quantum mechanics than heretofore; in these pages we shall derive and/or review all the quantum theory necessary for this purpose.

Our first example, the one-dimensional harmonic oscillator, can, like many similar models, be viewed in two contexts. At first let us solve it as the canonical model of the dynamics of a mass M tethered to the origin by a spring K whose Hamiltonian is,

$$H = \frac{p^2}{2M} + \frac{K}{2}x^2, \quad \text{where } p = \frac{\hbar}{i}\frac{\partial}{\partial x}. \quad (5.1)$$

The characteristic frequency is $\omega = \sqrt{K/M}$. Quantum theory requires $px - xp = \frac{\hbar}{i}$, i.e. $p = \frac{\hbar}{i}\partial/\partial x$. It also requires us to solve an eigenvalue problem for the energy eigenvalues E_n and for the corresponding wave functions $\Psi_n(x)$.

The Schrödinger equation $H\Psi_n(x) = E_n\Psi_n(x)$ is typically solved by series expansion, which yields Ψ_n as the product of an Hermite polynomial in x of degree n and of a Gaussian. The result: $E_n = \hbar\omega(n + 1/2)$, with $n = 0, 1, 2, \ldots$. The "zero point" energy is $\hbar\omega/2$. Here n just indicates the degree to which the system is excited — and *not* the number of particles occupying energy level $\hbar\omega$ as in the "number representation" to be introduced shortly.

There exists an efficient operator procedure to solve the same problem. It involves the two operators, a and its Hermitean conjugate a^+, respectively "lowering" and "raising" operators:

$$a = \frac{p - ix\sqrt{KM}}{(2\hbar\sqrt{KM})^{1/2}}, \quad \text{and} \quad a^+ = \frac{p + ix\sqrt{KM}}{(2\hbar\sqrt{KM})^{1/2}}, \tag{5.2}$$

normalized so as to satisfy $aa^+ - a^+a = 1$. Direct substitution of (5.2) into (5.3) and comparison with (5.1) shows that H can also be written as,

$$H = \hbar\omega(a^+a + 1/2). \tag{5.3}$$

It follows that $a^+a\Psi_n = n\Psi_n$ in general, with $n = 0, 1, 2, \ldots$, and $a^+a\Psi_0 = 0$ in the ground state. This last implies $a\Psi_0 = 0$, an homogeneous first-order differential equation for $\Psi_0(x)$. Its solution yields $A\exp{-x^2/d^2}$, where $d^2 = 2\hbar/\sqrt{KM}$ and A, the normalization constant, is adjusted such that $\int_{-\infty}^{\infty} dx|\Psi_0(x)|^2 = 1$. Once $\Psi_0(x)$ is normalized, the probability density for finding the particle at x in the ground state is $P(x) = |\Psi_0(x)|^2$.

Excited states are simply found by repeated applications of a^+ onto the ground state. Using only normalized Ψ's one finds,

$$\Psi_n(x) = \frac{1}{\sqrt{n!}}(a^+)^n\Psi_0(x), \quad \text{i.e. } \Psi_n(x) = \frac{1}{\sqrt{n}}a^+\Psi_{n-1}(x). \tag{5.4a}$$

Just as a^+ steps n up one integer at a time, a steps it down by precisely one integer:

$$a\Psi_n(x) = \sqrt{n}\Psi_{n-1}(x) \tag{5.4b}$$

Actually, all this information is summarized in the following *commutator* algebra:

$$aa^+ - a^+a \equiv [a, a^+] = 1, \quad aH - Ha \equiv [a, H] = \hbar\omega a, \quad \text{and} \quad [H, a^+] = \hbar\omega a^+. \tag{5.5}$$

Problem 5.1.   Prove the commutation relations in (5.5), using the definitions (5.1)–(5.3) for the various operators. Prove, then use the following identities to simplify the calculations:

1. $[a, AB] = A[a, B] + [a, A]B$, where a, A and B are arbitrary,
2. $[A, B] = -[B, A]$, and
3. $(AB)^+ = B^+ A^+$

   A^+ is denoted "Hermitean conjugate" of A if each is the transpose *and* the complex-conjugate of the other. The operators a and a^+ are Hermitean conjugates of one another, whereas H is its own Hermitean conjugate — i.e. it is "self-adjoint".

   After inverting Eq. (5.2) for p and x, expressing them as linear combinations of a and a^+,

4. *prove* x and p in this representation are both self-adjoint.

The *time-dependent* Schrödinger equation, $H\Phi(x,t) = i\hbar\partial\Phi(x,t)/\partial t$ can be solved in general using arbitrary coefficients c_n,

$$\Phi(x,t) = \sum_{n=0}^{\infty} c_n \Psi_n(x) e^{-itE_n/\hbar} . \qquad (5.6)$$

At $t = 0$, Φ is matched to initial conditions, subject to $\sum |c_n|^2 = 1$. Then the probability distribution $P(x,t) = |\Phi|^2$ describes the quantum-mechanical motion of the particle. If P is sharply peaked at a value $x(t)$ (this may require including states of sufficiently high n in the wavepacket (5.6),) it is found that the peak of the wavepacket at $x(t) \approx x_0 \cos \omega t$ emulates the harmonic motion of a classical particle. In summary, the Hamiltonian in (5.1) and its eigenstates describe every detail of the single normal mode.

Next we want to see the results of perturbing this Hamiltonian. As a solvable example, suppose the perturbation of the quadratic potential to be *linear.*[a]

$$H = \hbar\omega(a^+ a + 1/2) + g(a + a^+) \qquad (5.7)$$

The *shift transformation* preserves the commutation relations (5.5) and Hermitean conjugation: $a = b + f$ and $a^+ = b^+ + f^*$ where f is a

---

[a]If the perturbation were chosen differently, say as $H' = g'(a + a^+)^4$, H could not be diagonalized *explicitly* as in (5.8). Still, there *always does* exist a complete set of eigenfunctions Ψ_m (provided only that the perturbation H' is Hermitean). The low-lying eigenvalues E_m can, in fact, be calculated to arbitrary accuracy.

number and b an operator similar to a. It is therefore *unitary*. The new $H = \hbar\omega((b^+ + f^*)(b + f) + 1/2) + g(b + f + b^+ + f^*)$.

We choose the transformation parameter $f = -g/\hbar\omega$ to cancel all the linear terms. The final result $H = \hbar\omega(b^+b + 1/2) - g^2/\hbar\omega$ appears simple, i.e.

$$E_m(g) = \hbar\omega(m + 1/2) - g^2/\hbar\omega \,, \tag{5.8}$$

but *eigenstates* are strongly affected. For example expanding the old ground state Ψ_0 in the new Ψ'_m's yields: $e^{-\frac{f^2}{2}} \sum_{m=0}^{\infty} \frac{f^m}{\sqrt{m!}} \Psi'_m$. Next, we examine how to accommodate the many-body problem to the thermodynamic limit $N \to \infty$, starting with "free" particles (i.e. individual particles with no mutual interactions).

5.2.   Number Representation and the Many-Body Problem

Next let us suppose N identical, *noninteracting* particles "live" in, i.e. share, the same parabolic well. Their Hamiltonian is,

$$H = \sum_{i=1}^{N} H_i = \sum_{i=1}^{N} \left\{ \frac{p_i^2}{2m} + \frac{K}{2} x_i^2 \right\} . \tag{5.9}$$

Because they don't interact, their wave functions and probabilities should just multiply, i.e. $P(x_1, \ldots, x_N) = |\Psi_m(x_1)\Psi_{m'}(x_2) \ldots|^2$. In the event they are all bosons with the same mass and are subject to the same potential, P should be invariant under permutations. Clearly, the product is not symmetric unless all subscripts $m, m', \ldots$ are identical. The solution: symmetrize the over-all wave function $\Psi(x_1, \ldots, x_N)$ under permutations of the particles. Thus, let

$$\Psi(x_1, x_2, \ldots) = C \sum_P P\{\Psi_m(x_1)\Psi_{m'}(x_2) \ldots\} \tag{5.10}$$

where P stands for any one of the nontrivial permutations of the N coordinates and C is the requisite normalization factor.[b] The permutations are all "degenerate", i.e. share a common energy $E = \sum_{m=0}^{\infty} E_m(0)n_m$, with n_m the number of particles in the mth level and $E_m(0) = \hbar\omega(m + 1/2)$. From this we deduce the following feature of bosons, seemingly trivial but in fact

---

[b]If all the quantum numbers are different, there are $N!$ such permutations and $C = 1/\sqrt{N!}$. If all the quantum numbers are the same, there is just one and $C = 1$. In general C lies between these two extremes.

central to their statistical mechanics: if the particles are in fact totally indistinguishable, one is only entitled to know *how many* particles share m, how many share m', etc., in a given eigenstate, and not *which ones* are in m, in m', etc. Thus we can rewrite (5.9) in the following terms:

$$H = \sum_{m=0}^{\infty} E_m(0)c_m^+ c_m \tag{5.11}$$

where c_m and c_m^+ are new boson annihilation and creation operators in a new infinite-dimensional *occupation-number* space. In any acceptable eigenstate Ψ, we must have $\sum_{m=0}^{\infty} c_m^+ c_m \Psi = N\Psi$.

The harmonic oscillator analogy "works" very well for the c_m's. The generalization of the commutator algebra in Eq. (5.5) to these new operators is:

$$[c_m, c_{m'}^+] = \delta_{m,m'}, \quad [c_m, c_{m'}] = [c_{m'}^+, c_m^+] \equiv 0, \quad [c_m, c_m^+ c_m] = c_m. \tag{5.12}$$

The bosonic nature of the particles is mirrored by the fact that the creation operators commute with one another, i.e. $c_i^+ c_j^+ \Psi = c_j^+ c_i^+ \Psi$ for any state Ψ. Therefore one may not count permutations as new, linearly independent, states. This indistinguishability rule is what mandates Gibbs' factor $1/N!$ in the correspondence (classical) limit, as discussed previously. The harmonic oscillators underlying this *number representation* are *abstract*; there exist no corresponding dynamical coördinates p and x to which they refer. An independent harmonic oscillator mode is arbitrarily assigned to *each* one-particle level in the original problem. Dirac's notation labels their states in an intuitive manner, as follows:

$|0\rangle$ is the lowest occupation state, *alias* "the vacuum". Its Hermitean conjugate state is $\langle 0|$. Normalization is expressed through the relation $\langle 0|0\rangle = 1$. Normalized excited states

$$\Psi = |\ldots n_m \ldots\rangle = \prod_m \frac{(c_m^+)^{n_m}}{\sqrt{n_m!}}|0\rangle$$

are specified by the set of occupation numbers $\{n_m\}$. Because the operators c^+ commute they may be ordered in any way that is convenient. The Hermitean conjugate states are written as

$$\Psi^+ = \langle \ldots n_m \ldots| = \langle 0|\prod_m \frac{(c_m)^{n_m}}{\sqrt{n_m!}}.$$

5.3.　The Adiabatic Process and Conservation of Entropy

Let us pursue the example of noninteracting particles in a common potential. If the parameters of the harmonic well were changed there would be a new set of one-body energies $E_m(g)$. It would still be necessary to specify the new occupation numbers. Of course the total energy changes in the process, as $H \Rightarrow \sum_m E_m(g) c_m^+ c_m$. Suppose the initial eigenstate to be $|\ldots, n_m, \ldots\rangle$, labeled by the eigenvalues of the set of occupation numbers $\{c_m^+ c_m\}$, i.e. by the set $\{n_m\}$. If they remain unchanged as the parameter g is "turned on" we denote this a *quantum adiabatic transformation*. The *constancy of the occupation numbers* (if not of the individual energies nor even of the total energy) is what *defines* the quantum adiabatic process.[c]

At a given total energy E the entropy $\mathscr{S}(E)$ is defined proportional to the logarithm of the number of distinct configurations with energy E.

Let $W(R, N)$ be the number of ways in which $N = \sum n_m$ particles have energy $E = \hbar\omega(R + N/2)$, with $R = \sum m n_m$. Suppose $R = 3$. Without in any way distinguishing the N particles in the initial state, this value of R can only be achieved in the following 3 distinct modes:

1. $n_3 = 1$, $n_0 = N - 1$, all other n's $= 0$.
2. $n_2 = n_1 = 1$, $n_0 = N - 2$, all other n's $= 0$.
3. $n_1 = 3$, $n_0 = N - 3$, all other n's $= 0$.

Assuming each state 1,2,3 to have equal probability, we find: $W(3, N) = 3$ and $\mathscr{S}(E) = k \log 3$ for $E = \hbar\omega(3 + N/2)$ in this example.

By definition, in a quantum adiabatic transformation the occupation numbers in each of the states remain unchanged. Perturbing H as in Eq. (5.7) causes the energy of each member of the triplet to change to a new value $E \Rightarrow \hbar\omega(3 + N(1/2 - (g/\hbar\omega)^2))$. In this example the degeneracy of the triplet remains the same — hence so does the entropy $\mathscr{S}$. Here, *adiabatic* implies *isentropic*.

Problem 5.2.　Compute $W(R, N)$ for $R = N/10$. Call this $W_0(N)$. *Show* $\log W_0(N) \propto N$ in large N limit (Hint: modify the procedure of Sec. 1.2.)

---

[c]Regardless of appearances this constancy is not yet related to entropy; any boson state $|\ldots, n_m, \ldots\rangle$ is unique, hence has no entropy when considered by itself. A microscopic discussion of entropy has to start with a *collection* of such states, all of approximately the same energy.

But this special example of an isotropic harmonic well undergoing an adiabatic transformation *via* the linear shift analyzed above or, alternatively, through a gradual change in the spring constant or frequency $\omega(g)$, is almost unique — insofar as the transformation preserves all the individual multiplet structures — hence $\mathscr{S}$ as well. One other plausible case is that of the three-dimensional well with sides L_x, L_y, L_z changing by a common factor: $L_x, L_y, L_z \to \lambda L_x, \lambda L_y, \lambda L_z$. This too maintains initial symmetry and multiplet structure. But wherever the *quantum adiabatic process* changes the symmetry[d] it also affects the multiplet structure; the entropy must then be reëvaluated in a statistical sense and averaged over a finite energy range ΔE.[e] This challenge remains to be met.

But a greater challenge remains, which is to define the quantum adiabatic process for *interacting* particles — given that there is no set of well-defined individual quantum numbers to be preserved in this instance — let alone to identify in an isentropic process! We return to this topic ultimately, from a different perspective.

5.4. Many-Body Perturbations

As an alternative to making *implicit* changes in the underlying one-body Hamiltonian due to perturbing potentials affecting the individual E_m, we can introduce changes in the Hamiltonian *explicitly* using the field operators themselves. If the perturbation is in the *one-body* potentials the perturbing Hamiltonian has to be bilinear in the field operators, e.g.:

$$H_1 = g \sum_{m,m'} V_{m,m'} c_m^+ c_{m'} , \qquad (5.13)$$

where g is a coupling constant which is "turned on" from 0 to 1 and $V_{m,m'}$ is a given, known, matrix. Whatever the matrix elements $gV_{m,m'}$ may be, H_1 automatically conserves *total* number — even though it explicitly causes the occupation numbers of individual levels to change. In the many-body re-formulation of the solved example of Eqs. (5.7) and (5.8) one writes $H =$

---

[d]E.g. in the popular example of a piston adiabatically changing the length of a cylindrical cavity, the aspect ratio *does* change and the quantum adiabatic process is not obviously microscopically isentropic.

[e]Generally a quantum adiabatic process can only preserve the coarse-grained quantity $\mathscr{S}(\bar{E}, N)\Delta E$, with $\mathscr{S}$ averaged over a range of energies $E \pm \Delta E$, and not the microscopic function $\mathscr{S}(E, N)$ itself.

$H_0 + H_1$, where:

$$H_0 = \sum_m \hbar\omega \left(m + \frac{1}{2}\right) c_m^+ c_m , \quad H_1 = g \sum_{m=1}^{\infty} \sqrt{m}\{c_{m-1}^+ c_m + c_m^+ c_{m-1}\} .$$

$$(5.14)$$

Two-body interactions generically take the form:

$$H_2 = g \sum_{nmkl} V_{nmkl} c_n^+ c_m^+ c_k c_l . \tag{5.15}$$

Problem 5.3.   (A) Show the quadratic form $H_0 + H_1$ in (5.14) generalizes Eq. (5.7); exhibit the basis set for $N = 1, 2$ particles at $g = 0$. (B) The
perturbations above, H_1 and H_2, share an important property: for every factor c there is a factor c^+. It seems obvious that such perturbations explicitly
conserve the total number. However, the more standard way to prove that
$N_{op} \equiv \sum_{m=0}^{\infty} c_m^+ c_m$ is conserved is to show it commute with the Hamiltonian
and is thus a "constant of the motion". Therefore, *prove*: $[H_1, N_{op}] = 0$ and
repeat the calculation for H_2 given in (5.15).

5.5.   Photons

Photons are transverse excitations of the electromagnetic field polarized
perpendicular to the given direction of propagation (determined by the
orientation of the wavevector **k**.) The two polarizations of circularly polarized light are clockwise and counter-clockwise. (Alternatively, there exist
linearly polarized waves along two orthogonal, transverse, axes. Either way,
there are just two fields to quantize.) H takes the form,

$$H = \sum_{k,\alpha} \hbar c k a_{k,\alpha}^+ a_{k,\alpha} .$$

$\alpha = 1, 2$ labels the two possibilities (e.g. cw or ccw), and $a^+ a$ is the individual
modes' number operator, with eigenvalues $n_{k,\alpha} = 0, 1, 2, \ldots$ We can now
calculate Z to obtain $E(T)$, and $p(V, T)$ and other thermodynamic properties
of the photons "gas" of interest in astrophysics, optics, etc.

The $n_{k,\alpha}$ are the sole dynamical degrees of freedom, hence they are to
be summed after being weighed by the appropriate Boltzmann factor. As
the *number* of photons is not subject to any conservation law, there are
no extra constraints or Lagrange parameters such as a chemical potential.
(Neither the individual $n_{k,\alpha}$ nor their total number $\sum n_{k,\alpha}$ are prescribed.)

Then,

$$Z = Tr\{e^{-\beta H}\} = \prod_{k,\alpha} Z(\varepsilon_{k,\alpha}), \text{ where } \varepsilon_{k,\alpha} = \hbar c |\mathbf{k}| \geq 0,$$

independent of α, and each $Z(\varepsilon_{k,\alpha})$ takes the form $Z(\varepsilon)$,

$$Z(\varepsilon) = Tr\{e^{-\beta \varepsilon n}\} = \sum_{n=0}^{\infty} e^{-\beta \varepsilon n} = (1 - e^{-\beta \varepsilon})^{-1}. \tag{5.16}$$

The sum over occupation number n (short for $n_{k,\alpha}$) in Eq. (5.16) is identified as a standard, convergent, geometric series $(1 + x + x^2 \cdots) = (1-x)^{-1}$ that can be immediately evaluated as shown.

Using this result let us calculate the thermal-average occupancy,

$$\langle n_{k,\alpha} \rangle = \frac{0 + 1x + 2x^2 + \cdots}{1 + x + x^2 + x^3 + \cdots},$$

where $x = e^{-\beta \hbar c k}$. Eq. (4.4) suggests an easier way to get the same answer:[f]

$$\langle n_{\mathbf{k},\alpha} \rangle = -kT \frac{\partial}{\partial \varepsilon} \log Z(\varepsilon) = \frac{x}{1-x} = \frac{1}{e^{\beta \hbar c k} - 1} \tag{5.17}$$

(basically, Planck's law). At low energy compared with kT, $\langle n \rangle$ reduces to the classical equipartition result plus a correction eliminating the quantum zero-point motion, viz.

$$\langle n_{\mathbf{k},\alpha} \rangle = \frac{1}{e^{\beta \hbar c k} - 1} \xrightarrow[\beta \hbar c k \to 0]{} \frac{kT}{(\hbar c k)} - \frac{1}{2}. \tag{5.18}$$

It remains to calculate $F = -kT \log Z$, a different kind of sum. Because $Z = \prod_{k,\alpha} Z(\varepsilon_{k,\alpha})$, a product, its logarithm is a sum over allowed $\mathbf{k}$'s:

$$F = 2kT \sum_{k} \log\{1 - e^{-\beta \hbar c k}\} = 2VkT \int d\varepsilon \rho(\varepsilon) \log\{1 - e^{-\beta \varepsilon}\} \tag{5.19a}$$

with the factor 2 for the two polarizations α. The internal energy is

$$E = \frac{\partial(\beta F)}{\partial \beta} = 2 \sum_{k} \frac{\hbar c k}{e^{\beta \hbar c k} - 1} = 2V \int d\varepsilon \rho(\varepsilon) \frac{\varepsilon}{e^{\beta \varepsilon} - 1} \tag{5.20a}$$

Here we find it convenient to introduce the kinematic concept of *density of states* (dos) $\rho(\varepsilon)$, to take into account the manner in which the $\mathbf{k}$'s are selected. The following derivation explains how this is done.

---

[f]Treating ε as the coupling constant.

In a box L^3 the component $k_x = 2\pi m_x/L$, with $m_x = 0, \pm 1, \pm 2, \ldots$ and similarly with k_y and k_z. The m's form a *simple-cubic* 3D grid $\mathbf{m} \equiv (m_x, m_y, m_z)$ with unit lattice parameter. The Riemann sum $\sum_k$ is over m_x, m_y, and m_z and reduces, in the large volume limit, to a three-dimensional Riemann integral.

In summing any continuous function of the photons' energy $\varepsilon = \hbar c k$, say $\Phi(\hbar ck)$, one should always proceed in the same manner:

$$\sum_k \Phi(\hbar ck) \Rightarrow \iiint dm_x dm_y dm_z \Phi(\hbar ck) = \left(\frac{L}{2\pi}\right)^3 \int d^3 k \Phi(\hbar ck)$$

$$= \frac{V}{(2\pi)^3}(\hbar c)^{-3} 4\pi \int d\varepsilon\, \varepsilon^2 \Phi(\varepsilon) \quad (\text{using } k = \varepsilon/\hbar c)$$

$$\equiv V \int d\varepsilon\, \rho(\varepsilon) \Phi(\varepsilon) \tag{5.21}$$

Comparison of the third line with the second defines the *dos* $\rho(\varepsilon)$:

$$[\text{for photons}] \quad \rho(\varepsilon) \equiv \frac{4\pi}{(2\pi\hbar c)^3}\varepsilon^2 \tag{5.22}$$

When used in (5.19a) this leads to:

$$F = 2VkT\frac{4\pi}{(2\pi\hbar c)^3}\int d\varepsilon\, \varepsilon^2 \log\{1 - e^{-\beta\varepsilon}\}, \tag{5.19b}$$

hence

$$E = \frac{\partial(\beta F)}{\partial \beta} = 2V\frac{4\pi}{(2\pi\hbar c)^3}\int d\varepsilon\, \varepsilon^3 \frac{1}{e^{\beta\varepsilon} - 1}. \tag{5.20b}$$

According to Eq. (3.10) the radiation pressure on the walls of the box V has to be calculated isentropically. Following the discussion in Sec. 5.3 we shall evaluate it *adiabatically*, by keeping the number of photons at each $\mathbf{m}$ explicitly constant as the volume is changed — all the while maintaining spatial isotropy. Following this prescription one writes:

$$p = -\left.\frac{\partial E}{\partial V}\right|_{N,\mathscr{S}} = -\sum_{\mathbf{m},\alpha}\left[\frac{\partial}{3L^2\partial L}\left(\frac{\hbar c 2\pi\sqrt{m_x^2 + m_y^2 + m_z^2}}{L}\right)\right]\langle n_{k,\alpha}\rangle_{TA}$$

$$= \frac{E}{3V} \tag{5.23}$$

The alternative in (3.10) uses Eq. (5.19a):

$$p = \frac{\partial}{3L^2 \partial L}\left[2kT\sum_{\mathbf{m}}\log\{1 - e^{-\beta\hbar c 2\pi|\mathbf{m}|/L}\}\right]_T \equiv (1/3)E/V \,,$$

and yields an identical result, as the reader will wish to verify.

E as given in the form (5.20b) can be calculated in closed form. With $\beta\varepsilon \equiv x$ a dummy of integration, $E = 2V\frac{4\pi(kT)^4}{(2\pi\hbar c)^3}\int_0^\infty dx\frac{x^3}{e^x-1}$. This integral is I_3, a special case of I_n:

$$I_n = \int_0^\infty dx\,\frac{x^n}{e^x - 1} = \int_0^\infty dx\,x^n[e^{-x} + e^{-2x} + e^{-3x} + \cdots]$$

$$= \left[\int_0^\infty dx\,x^n e^{-x}\right] \times \left[1 + \frac{1}{2^{n+1}} + \frac{1}{3^{n+1}} + \cdots\right]$$

$$= \Gamma(n+1) \times \zeta(n+1) \tag{5.24}$$

with $\Gamma(x)$ the gamma function (cf. Chapter 1) and $\zeta(x) = \sum_{m=1} m^{-x}$ the Riemann zeta function, another function that frequently comes up in statistical mechanics. Both are extensively tabulated. In particular, $\Gamma(4) = 3!$ and $\zeta(4) = \pi^4/90$, i.e. $I_3 = \pi^4/15$.

Max Planck was awarded the 1918 Nobel prize in physics for obtaining the above results and extracting from them the following expressions for the energy and radiation pressure,

$$E_{\text{rad}} = V\frac{\pi^2}{15(\hbar c)^3}(kT)^4 \quad \text{and} \quad p_{\text{rad}} = \frac{\pi^2}{45(\hbar c)^3}(kT)^4\,. \tag{5.20c}$$

Stefan's 1879 law, Eq. (5.20c), as these formulas were originally known, summarized observations on emitted energy and radiation pressure from earthly and astronomical objects in experiments dating as far back as the mid-19th Century. Despite an earlier heuristic classical derivation by Boltzmann, this law is, in fact, incompatible with classical thermodynamics of the electromagnetic spectrum. It is therefore only historical justice that this work earned Planck the sobriquet, "father of the quantum".

5.6. Phonons

The speed of sound of long wavelength transverse vibrational modes (quantized as "phonons") in solids is related to the shear modulus, that of the longitudinal modes to the bulk modulus. These are the so-called "acoustic modes". In solids with a *basis*, that is, whose unit cell contains two or more ions, internal oscillations of each cell produces distinct, "optical", modes. As indicated by this nomenclature, their oscillating dipole moment $qx(t)$ readily emits or absorbs photons, typically in the infrared.

In a solid consisting of N unit cells of z atoms each, there are $3Nz$ normal modes. Of these, $2N$ are the transverse "acoustical phonons" which propagate at the speed of sound s (all z atoms in each cell remaining more or less in phase), and N are longitudinal phonons (compressional waves) propagating at a somewhat higher speed s_l. The remainder constitute the "optical phonon" spectrum, based on intra-cell oscillations; as the dispersion of optical modes is of no great import one often ignores it altogether and assigns to the optical modes a common frequency ω_0. Similarly, statistical mechanical calculations are simpler if one treats the three acoustic modes on a common basis, using a suitably averaged speed of sound.[g]

In the early 20th Century, Albert Einstein and Peter Debye proposed competing theories for the vanishing of the heat capacity in non-metallic solids at low temperatures, both based on Planck's formulation for the photons. As it turned out, both were right. Whereas Einstein's theory applies to optical phonons, wherever they exist — such as in *NaCl* — Debye's theory applies to the acoustical phonons present in all solids.

We examine the Einstein theory first, as it is the simpler. One supposes there are N_E distinct normal modes having a common frequency ω_E, with N_E some integer multiple of the number of cells, N. Neglecting nonlinearities, interactions among normal modes, scattering etc., the relevant partition function is simply:

$$Z_E = \left(\frac{1}{1 - e^{-\beta\hbar\omega_E}} \right)^{N_E} . \tag{5.25}$$

The internal energy is,

$$E_E = N_E\hbar\omega_E \frac{1}{e^{\beta\hbar\omega_E} - 1} . \tag{5.26}$$

Corresponding to a heat capacity,

$$C_E(T) = kN_E[x(T)/\sinh x(T)]^2 , \quad \text{where } x(T) \equiv \hbar\omega_E/2kT . \tag{5.27}$$

At high temperatures $x \to 0$ and the heat capacity attains its classical limit, the *Dulong–Petit* law: k *per* harmonic oscillator. At low temperatures $x \gg 1$

---

[g]There is no space here for deriving these standard results and the interested reader is encouraged to delve into books on the solid state or more specialized treatises such as J. Ziman's *Electrons and Phonons* (Oxford Univ. Press, London, 1960).

and the heat capacity vanishes, although when compared to observation it is observed that the exponential decrease $C \propto \exp -\hbar\omega_E/kT$ is far too abrupt.

Debye's theory reconciles theory to experiment. Let $N_D = 3N$ normal modes have dispersion $\omega(k) = sk$, or energy $e(k) = \hbar sk$, where the speed of sound $s \propto (\text{elastic constant}/\text{atomic mass})^{1/2}$. The maximum k denoted k_D, is the radius of the "Debye sphere", chosen to have the same volume as the first Brillouin Zone. Therefore, like the latter, it contains precisely $N = (L/a_0)^3$ points:

$$\frac{4}{3}\pi \left(\frac{k_D L}{2\pi}\right)^3 = N \Rightarrow k_D = (3/2)^{1/3}(2\pi)^{2/3}/a_0 = 3.89778/a_0 \,. \qquad (5.28a)$$

The cutoff $\varepsilon_D = \hbar s k_D \equiv k\theta_D$ is also $\propto (\text{elastic constant}/\text{atomic mass})^{1/2}$. We can obtain the dos of acoustic phonons without additional calculation by comparing this with Eq. (5.22) for the photons, setting $\rho(\varepsilon) = A\varepsilon^2$ for $\varepsilon < \varepsilon_D$ and $\rho = 0$ for $\varepsilon \geq \varepsilon_D$. Assuming the speeds of longitudinal and transverse modes are equal and using the definition of the cutoff,

$$A \int_0^{k\theta_D} d\varepsilon\varepsilon^2 = N \Rightarrow A = 3N(k\theta_D)^{-3} \,. \qquad (5.28b)$$

The total energy of the 3 branches of acoustic phonons is,

$$E_D = 3\left\{A \int_0^{k\theta_D} d\varepsilon\varepsilon^3 \frac{1}{e^{\beta\varepsilon} - 1}\right\} = 9N(kT)x^{-3} \int_0^x dt\, t^3 \frac{1}{e^t - 1} \qquad (5.29)$$

where $x \equiv (\theta_D/T)$. At low temperatures, x is large and can be taken to infinity. Then — except for the trivial replacement of s by c and for a numerical factor $2/3$ — Eq. (5.29) reduces to the analogous expressions for photons, Eqs. (5.20a–c). Explicitly: as $T \to 0$, $E_D \to N(9\pi^4/15)(kT)^4/(k\theta_D)^3$. Thus, the heat capacity vanishes as T^3 at low temperatures, far less abruptly than the exponential decrease of the optical phonons. In insulators the value of θ_D is determined experimentally by fitting the low temperature C_v to this result. In metals the far greater heat capacity of electrons masks the Debye terms.

In the high-temperature limit $x \to 0$. The integral (5.29) reduces to $x^3/3$. Then $E_D \to 3NkT$ and C attains its Dulong–Petit "classical" limit $3Nk$. To examine E_D at intermediate temperatures, or to determine the rate of approach to either limit, one requires the Debye integral[h] D_3.

---

[h]Defined and tabulated in NBS Tables, Abramowitz and Stegun, Eds.

Other Debye integrals D_n have proved useful as well. In general they are defined as follows:

$$D_n(x) = \int_0^x dt\, t^n \frac{1}{e^t - 1} = \Gamma(n+1)\zeta(n+1)$$
$$- \sum_{k=1}^{\infty} e^{-kx} \left\{ \frac{x_n}{k} + n\frac{x^{n-1}}{k^2} + n(n-1)\frac{x^{n-2}}{k^3} + \cdots + n!\frac{1}{k^{n+1}} \right\}. \quad (5.30a)$$

A different series expansion converges more efficiently at small x:

$$D_n(x) = x^n \left\{ \frac{1}{n} - \frac{x}{2(n+1)} + \sum_{k=1}^{\infty} \frac{B_{2k} x^{2k}}{(2k+n)(2k)!} \right\} \quad (5.30b)$$

where B_n are the *Bernoulli numbers*.

The generating function of the Bernoulli numbers is

$$\frac{t}{e^t - 1} = \sum_{n=0}^{\infty} B_n \frac{t^n}{n!}, \quad \text{i.e. } B_0 = 1, B_1 = -1/2, B_2 = 1/6,$$

$$B_4 = -1/30, B_6 = 1/42, B_8 = -1/30, B_{10} = 5/66, \text{etc.}$$

Odd-subscripted B's beyond B_1 all vanish.

Whenever extraordinary accuracy is required (beyond the Debye theory) one may have recourse to the exact *dos*, $\rho(\varepsilon)$ defined as follows: $\frac{1}{N}\sum_{k,\alpha}\delta(\varepsilon - \varepsilon_\alpha(k))$ for arbitrary dispersion $\varepsilon_\alpha(k)$, and evaluated numerically. The resulting corrections to the Debye theory are typically most pronounced in the vicinity of $T/\theta_D = 1$, and negligible elsewhere.

5.7.   Ferromagnons

Consider a fully polarized ferromagnet of spins S with its magnetization precisely along the z-axis. Its magnetic moment is $M = NS$. An excitation, denoted ϕ_j, is constructed out of a lowered spin $S_z = S - 1$ on the jth site (with S_z remaining $= S$ on all the other sites.) Because the choice of site is arbitrary, the eigenstate ("magnon") is a plane-wave linear combination $\propto \sum \phi_j e^{ik \cdot Rj}$. One calculates the energy of this eigenstate as $\varepsilon(k) = JS(ka_0)^2$

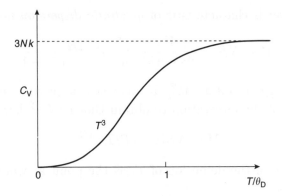

Fig. 5.1.   Heat capacity of a Debye insulating solid.
$C(T/\theta_D)/Nk_B$, calculated with the aid of Eqs. (5.29)*ff*, is a universal function of T/θ_D.

at long wavelengths, with J the so-called exchange parameter.[i] The Brillouin zone containing N points is again approximated by a Debye sphere of radius k_D as given in Eq. (5.28a).

With each magnon decreasing the magnetization by one unit, the Curie temperature is obtained as that temperature T_c at which NS magnons are thermally generated, i.e. at which $M \to 0$. Once again the density of states (*dos*) is helpful to obtain the answer. However, because the dispersion is now quadratic, we can no longer use the formulas derived for photons and phonons with linear dispersion. We derive expressions appropriate to quadratic dispersion as follows: setting $\varepsilon \equiv Ak^2$ (with $A \equiv Ja_0^2S$ and a_0 the lattice parameter), an arbitrary sum takes the form,

$$\sum_k \Phi(Ak^2) = \left(\frac{L}{2\pi}\right)^3 \int d^3k\Phi(Ak^2)$$

$$= \frac{V}{(2\pi)^3}\frac{2\pi}{A^{3/2}}\int d\varepsilon\varepsilon^{1/2}\Phi(\varepsilon) \equiv V\int d\varepsilon\rho(\varepsilon)\Phi(\varepsilon).$$

Comparison of the last two terms yields $\rho(\varepsilon)$,

$$\rho(\varepsilon) = (2\pi)^{-2}A^{-3/2}\varepsilon^{1/2}. \qquad (5.31)$$

---

[i]See D. C. Mattis, *The Theory of Magnetism* I, Springer, Berlin, 1981, Chap. 5, or *The Theory of Magnetism Made Simple*, World Scientific, forthcoming.

The 1/2 power is characteristic of *quadratic dispersion in* 3D. Then,

$$M = NS - \frac{V}{(2\pi)^2 A^{3/2}} \int_0^{\varepsilon_m} d\varepsilon \varepsilon^{1/2} \frac{1}{e^{\beta \varepsilon} - 1}, \qquad (5.32)$$

with ε_m the energy cut-off at Ak_D^2. If this last is approximated by ∞, β can be factored out of the integration to obtain Bloch's $T^{3/2}$ law:

$$M \approx NS[1 - (T/T_c)^{3/2}], \qquad (5.33)$$

upon making use of the definition of T_c as the point at which spontaneous *LRO* disappears.

Although it proves useful at low temperatures, Bloch's law fails at temperatures comparable to T_c owing to several sources of error:

- a finite cutoff ε_m cannot be approximated by ∞ near T_c; (5.33) is therefore a lower bound to M as given by the "Debye integral" in (5.32).
- the dispersion is not well approximated by k^2 at large k, and finally,
- the existence of non-negligible magnon-magnon interactions.

The last are particularly significant in *low dimensions*, where they cause multi-magnon bound states (and other complexes) to be created at low energies. These are capable of destroying the long-range order at all finite temperature, effectively eliminating the Curie transition by pushing it down to $T_c = 0$. But in $d \geq 3$ dimensions the bound states — if any — exist only at finite energy and their low-T contributions are exponentially small. Also, according to calculations by F. J. Dyson, the correction to the magnetization in 3D due to such *magnon-magnon* scattering is only $O(T/T_c)^4 \ll O(T/T_c)^{3/2}$, i.e. insignificant at low $T \ll T_c$.

Problem 5.4.   Generalizing Eqs. (5.32) and (5.33), find kT_c and the correct form of $M(T/T_c)$ in 3D, 4D,..., dD,..., as a function of JS, for $S = 1/2, 1, \ldots, \infty$.

Problem 5.5.   In the *antiferromagnet*, two interpenetrating sublattices have *oppositely* oriented magnetizations. The effect of a magnon is to decrease magnetization in each sublattice by one unit of angular momentum. Given that the dispersion for "antiferromagnons" is $\varepsilon(k) = JSa_0|k|$, i.e. linear, derive the law for the low-temperature sublattice order parameter σ (Note: $\sigma \equiv 1$ if the sublattices are perfectly ordered and $\sigma = 0$ at the Néel point T_N).

5.8. Conserved Bosons and the Ideal Bose Gas

Atoms having even numbers of electrons and nucleons qualify as "conserved" bosons, although the formation of diatomic or multi-atomic molecules (bound states) inordinately complicates matters, as they must be considered separate species from the point of view of the statistical theory. With its closed shells and weak van der Waals' attraction, and its light mass, helium (^{4}He) is incapable of forming diatomic molecules and comes closest to the ideal. We'll consider the properties of a perfect gas of bosons first and derive the theory of the so-called *ideal bose* gas condensation, a strange phenomenon that occurs only in the *absence* of two-body forces.

In later sections we examine the effects of dimensionality and of one- and two-body forces, and then touch upon the theories of liquid, superfluid, helium. Here we shall assume a 3D volume V containing N non-interacting bosons. Fixing the number of particles at N in the volume V requires (as it did in Eq. (4.13)) the introduction of a chemical potential $\mu(T)$ into the individual energies. We add and subtract μN into the Hamiltonian:

$$H = \sum_k \left(\frac{\hbar^2 k^2}{2M} - \mu \right) c_k^+ c_k + \mu N . \tag{5.34}$$

With $\varepsilon(k) \equiv \hbar^2 k^2 / 2M$, the constraint takes the following form:

$$N = \sum_k \frac{1}{e^{\beta(\varepsilon(k) - \mu)} - 1} = V \int_0^\infty d\varepsilon \rho(\varepsilon) \frac{1}{e^{\beta(\varepsilon - \mu)} - 1} , \tag{5.35a}$$

$\rho(\varepsilon)$ being given in (5.31) as the dispersion is quadratic, with $A = \hbar^2/2M$. This yields $\mu(N/V, T)$. [Exercise for the reader: verify that Eq. (5.35a) is equivalent to $\partial F / \partial \mu = 0$.] The requirement that $\bar{n}_k = (\exp \beta(\varepsilon(k) - \mu) - 1)^{-1} \geq 0$ for all k including $k \to 0$ obliges μ to be negative. As T is decreased, μ, the solution to (5.35a), approaches the axis and effectively vanishes at a temperature T_c, as seen in the figure on p. 101.

Let us reëxamine this equation carefully. Above T_c it reads:

$$N = V(2\pi\hbar)^{-3} M^{3/2} 4\pi\sqrt{2} \int_0^\infty d\varepsilon \varepsilon^{1/2} \frac{1}{e^{\beta(\varepsilon - \mu)} - 1} \tag{5.35b}$$

But *at* T_c precisely, where $\mu \to 0$, the integral simplifies:

$$N = V(2\pi\hbar)^{-3} M^{3/2} 4\pi\sqrt{2} (kT_c)^{3/2} I_{1/2} \tag{5.35c}$$

According to Eq. (5.24), $I_{1/2} = \Gamma(3/2)\zeta(3/2)$, with $\Gamma(3/2) = \sqrt{\pi}/2$ and $\zeta(3/2) = 2.612\ldots$. Then, $N/V = 2.612(MkT_c/(2\pi\hbar^2))^{3/2}$. Defining $r^3 = V/N$ as the volume *per* particle (the "specific volume") we solve for kT_c as a function of r:

$$kT_c = 3.31 \left(\frac{\hbar^2}{Mr^2} \right) . \tag{5.36}$$

Below T_c we consider the occupation of momentum $k = 0$ explicitly:

$$\bar{n}_0 = (\exp \beta(0 - \mu) - 1)^{-1} \approx -kT/\mu + \frac{1}{2} + O(\mu/kT) + \cdots \tag{5.37}$$

If μ is not strictly zero but is $O(1/L^3)$ this occupation number can become macroscopic, $O(L^3)$. That feature is the essence of *Bose-Einstein condensation*. What about the occupancy of the next lowest levels in the hierarchy, say at $\varepsilon_1 = (2\pi)^2 \hbar^2/2ML^2$, in this temperature range? First, note $\varepsilon_1 \gg |\mu|$, hence μ drops out and

$$\bar{n}_1 = (\exp \beta(\varepsilon_1 - \mu) - 1)^{-1} \approx kT/\varepsilon_1 + \frac{1}{2} + O(\varepsilon_1/kT) + \cdots \propto L^2(\cdots) \tag{5.38}$$

As this is only $O(L^2)\bar{n}_1$ is *sub*macroscopic in 3D, hence of no consequence in the thermodynamic limit. (Of course, this argument requires reëxamination in 2D and even more so in 1D!)

Adding the occupation of the discrete level at $k = 0$ to that of the remaining levels (still expressed as an integral,) we now find for the conservation law in the low-temperature range $T < T_c$:

$$N = \bar{n}_0 + V(2\pi\hbar)^{-3}M^{3/2}4\pi\sqrt{2} \int_0^\infty d\varepsilon \varepsilon^{1/2} \frac{1}{e^{\beta\varepsilon} - 1}$$

$$= \bar{n}_0 + V(2\pi\hbar)^{-3}M^{3/2}4\pi\sqrt{2}(kT)^{3/2}I_{1/2}$$

$$= \bar{n}_0 + N(T/T_c)^{3/2} \tag{5.35d}$$

upon using (5.35c) explicitly to eliminate the constants and other parameters.

As it was for the magnetization, the result is again a $T^{3/2}$ law for the order parameter $\sigma \equiv \bar{n}_0/N$:

$$\sigma = [1 - (T/T_c)^{3/2}] . \tag{5.39}$$

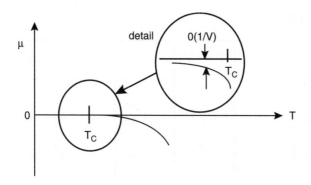

Fig. 5.2. Schematic plot of μ versus T near T_c.

In 3D, at T_c the chemical potential vanishes: $\mu \to 0$. However, $V \cdot \mu \neq 0$, as indicated in the detail.

Above T_c we simplify (5.35b) by using (5.35c) to eliminate parameters, obtaining the following implicit equation for $\mu(T)$:[j]

$$\int_0^\infty dx \, x^{1/2} \frac{1}{\xi(T)e^x - 1} = (T_c/T)^{3/2} I_{1/2} . \qquad (5.40)$$

The bose fluid and the ideal ferromagnet, two physically distinct systems, satisfy similar sets of equations: Eqs. (5.35)–(5.39) versus Eqs. (5.32)–(5.33). The existence of such a correspondence had been previously noted in connnection with the lattice gas. $\xi(T)$ given in (5.40), plotted in Fig. 5.3 for $T > T_c$, complements Fig. 5.2. Next, consider $d\xi/dT$ near T_c.

Differentiation of both sides of Eq. (5.40) *w.r.* to ξ at T_c yields $\partial T/\partial \xi|_c \times$ finite quantity on the right and a divergent integral on the left. Consequently $\partial T/\partial \xi|_c = \infty$ (this can be barely discerned in the figure.) We thus establish $\partial \mu/\partial T|_c = 0$. Because $\mu = 0$ for $0 < T < T_c$, we now know that both μ and $\partial \mu/\partial T$ are continuous across the phase transition. Is then this phase transition second-order? Before deciding one must examine this rather paradoxical fluid in greater detail. As the first and crucial step let us next examine the internal energy in the fluid, $E(T)$.

5.9. Nature of "Ideal" Bose–Einstein Condensation

The internal energy of the ideal bose gas is given by distinct expressions below and above T_c although in principle, it is always just $E(T) =$

---

[j]Or rather, for the more convenient quantity $\xi \equiv \exp(-\mu/kT) \leq 1$, a.k.a. "fugacity".

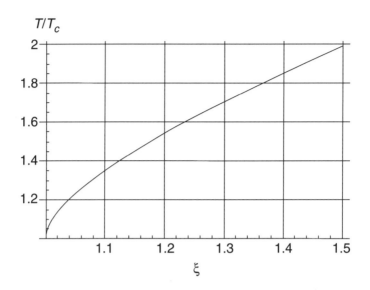

Fig. 5.3. (T/T_c) versus $\xi = \exp{-\mu/kT}$ for $T > T_c$.

Numerical solution of Eq. (5.40) in the temperature interval T_c to $2T_c$.

$\sum_k \varepsilon(k)\frac{1}{\xi(T)e^{\beta\varepsilon(k)}-1}$, with $\varepsilon(k) = \hbar^2 k^2/2M$ and $\xi = 1$ for $T < T_c$. Above T_c, $\xi(T)$ is given by the solution of Eq. (5.40) as illustrated in the above figure.

Below T_c we can use Eq. (5.35c) to eliminate various constants and to show that $E \propto N$, i.e. that it is properly extensive. After some straightforward algebra one obtains expressions involving gamma and zeta functions at low temperatures:

$$E(T) = NkT(T/T_c)^{3/2}\frac{\Gamma(5/2)\zeta(5/2)}{\Gamma(3/2)\zeta(3/2)} \quad \text{for } T \le T_c \qquad (5.41a)$$

and above T_c, involving $\xi(T)$ as well:

$$E(T) = NkT(T/T_c)^{3/2}\frac{\int_0^\infty dt\, t^{3/2}\frac{1}{\xi(T)e^t-1}}{\Gamma(3/2)\zeta(3/2)} \quad \text{for } T > T_c. \qquad (5.41b)$$

At low temperature $C_v \propto T^{3/2}$. At high temperature $C_v(T) \to 3/2Nk$, the Dulong–Petit limit for this model.[k] Because $\partial\mu/\partial T$ and $\partial\xi/\partial T$ vanish at T_c, C_v is continuous through the phase transition. It does however peak at

[k]Note: This is half the high-T limiting value for a solid of N atoms, a reasonable enough result given the absence of potential energy in the ideal gas. However, $T^{3/2}$ is not a good fit to the low-temperature heat capacity of liquid Helium.

a cusp. Differentiating (5.41a) we obtain the value at the apex at T_c:

$$C_v(T_c) = \left(\frac{5}{2}\right)\left(\frac{3}{2}\right)\frac{\zeta(5/2)}{\zeta(3/2)}Nk = \left(\frac{15}{4}\right)\left(\frac{1.341}{2.612}\right)Nk = 1.925Nk\,. \quad (5.42)$$

Although a cusp in the heat capacity seemingly indicates a *third-order* discontinuity in F, in this instance it does not unambiguously determine the nature of the phase transition. Let us examine the equation of state. The pressure is obtained, much as it was for the photon gas in Eq. (5.23), by the use of Eq. (3.10):

$$p = -\sum_{\mathbf{m}}\left[\frac{\partial}{3L^2\partial L}\left(\frac{2\pi\sqrt{m_x^2 + m_y^2 + m_z^2}}{L}\right)^2 \hbar^2/2M\right]\langle n_k\rangle_{TA}$$

$$= \frac{2E}{3V} \quad\quad\quad\quad\quad\quad\quad\quad\quad\quad\quad\quad\quad\quad\quad\quad (5.43)$$

Then,

$$p = \frac{2}{3r^3}kT(T/T_c)^{3/2}\frac{\Gamma(5/2)\zeta(5/2)}{\Gamma(3/2)\zeta(3/2)}$$

below T_c.

Elimination of T_c with the aid of (5.36) simultaneously eliminates r from the formula and produces a most unusual low-temperature equation of state:

$$p = \gamma T^{5/2} \quad \text{for } T \leq T_c\,, \quad\quad\quad\quad (5.44)$$

in which γ is a lumped constant independent of N, V or T. Because $\partial p/\partial V = 0$, the isothermal compressibility, Eq. (3.14a), diverges *everywhere* along the coëxistence curve for $T \leq T_c$:

$$\kappa_T = -\frac{1}{V}\frac{\partial V}{\partial p}\bigg|_T \to \infty \quad\quad\quad\quad (5.45)$$

as might be expected in a regime where there are two phases (the condensed portion at $k = 0$ being the "superfluid" phase, and the rest the "normal" phase) in thermodynamic equilibrium. Such a two-phase regime argues for a *first-order* phase transition similar to the liquid-vapor transition treated in an earlier chapter, and identifies T_c with the critical point.

When (5.45) is inserted into Eq. (3.13) we find $C_p \to \infty$ *in the entire temperature range $T < T_c$*. This is an absurdity that counts as one of the many paradoxes of the "ideal" bose gas phenomenon. Such inconsistencies can only be cured by introducing interactions — however weak they might be — among the constituent particles. But before doing so, we investigate the effects of dimensionality.

5.10.   Ideal Bose-Einstein Condensation in Low Dimensions

It is generally (but, as we shall see, erroneously) believed that Bose–Einstein condensation occurs only in dimensions $d \geq 3$. The argument, detailed in the following paragraph, is based on number conservation and on the peculiarities of the *dos* $\rho(\varepsilon)$ at the band edge ($\varepsilon \to 0$).

Problem 5.6.   Show that for free particles in d dimensions, with $\varepsilon \propto k^2$, $\rho(\varepsilon) = A_d \varepsilon^{-1+d/2}$ for $\varepsilon > 0$ and is $\equiv 0$ for $\varepsilon < 0$. Determine A_d for $d = 1, 2, 3$.

Insert $\rho(\varepsilon) = M/2\pi\hbar^2 = $ constant (for $\varepsilon > 0$) in 2D and $\rho(\varepsilon) = \sqrt{(M/2\pi^2\hbar^2)} \times \varepsilon^{-1/2}$ in 1D (for $\varepsilon > 0$) (cf. Problem 5.6) into Eqs. (5.35). The important feature to notice is that in $d > 2$ dimensions $\rho(\varepsilon)$ vanishes at the band edge, $\varepsilon \to 0$, whereas for $d \leq 2$ it does not.

Next define the interparticle separation parameter r in terms of particle density: in 2D, $N/L^2 = 1/r^2$ while in 1D, $N/L = 1/r$. To obtain the critical temperature, set $\mu = 0$ and modify Eq. (5.35c) for the lower dimensions by using the appropriate dos. It yields:

$$N = L^2 \frac{M}{2\pi\hbar^2}(kT_c)I_0\,, \ \text{i.e.} \ kT_c = \frac{2\pi\hbar^2}{Mr^2I_0} \quad \text{in 2D}\,, $$

and similarly,

$$N = L\sqrt{\frac{M}{2\pi^2\hbar^2}}(kT_c)^{1/2}, \quad kT_c = \frac{2\pi^2\hbar^2}{Mr^2I_{-1/2}} \quad \text{in 1D}\,. $$

$$(5.46)$$

The relevant integrals, I_0 in 2D and $I_{-1/2}$ in 1D, special cases of the integral defined in Eq. (5.24), are both manifestly divergent. Hence $T_c = 0$ in both instances. This might seem to ensure that at finite T in low dimensions the fluid is always in its high-temperature, normal, phase. However, the very same feature that causes I_n to diverge for $n \leq 0$ also renders the system unstable against arbitrarily weak random perturbations and promotes the appearance of bound states at energies $\varepsilon \leq 0$ in $d < 3$ dimensions. Whether these bound states are isolated or form an "impurity band", their presence completely alters the situation.

This feature can be demonstrated just by introducing a single weak, short-ranged, potential well at some fixed, random, position in 2D. If sufficiently short-ranged it will sustain only a single bound state with finite binding energy. The modified dos is

$$\rho(\varepsilon) = (1/L^2)\delta(\varepsilon + \Delta) + M/2\pi\hbar^2\theta(\varepsilon)\,, \tag{5.47}$$

where Δ is the binding energy and $\theta(x)$ is the unit step ($\theta(x) = 1$ for $x > 0$ and 0 for $x < 0$.) Now μ can never exceed $-\Delta$, and if there is a nonvanishing T_c, this will have to be determined using Eq. (5.35) after setting $\mu = -\Delta$:

$$N = L^2(2\pi\hbar^2)^{-1}M \int_0^\infty d\varepsilon \frac{1}{e^{\beta_c(\varepsilon+\Delta)} - 1}$$

$$= L^2(2\pi\hbar^2)^{-1}MkT_c \int_0^\infty dx \frac{1}{e^{\beta_c\Delta}e^x - 1}\,. \tag{5.48a}$$

This integral can be evaluated in closed form. The result is an implicit equation for T_c,

$$\frac{2\pi\hbar^2}{Mr^2} = kT_c \log\left(\frac{1}{1 - e^{-\Delta/kT_c}}\right)\,. \tag{5.48b}$$

For $\Delta > 0$ this equation *always has a solution*. The *rhs* increases monotonically with T_c from 0 to ∞, allowing for a unique solution $T_c(r)$ at any given value of r.

Below T_c the number of particles N_Δ condensed into the eigenstate at $\varepsilon = -\Delta$ is found by straightforward application of Eq. (5.35d) to the present example. It is,

$$N_\Delta = N\left(1 - \frac{T\log(1 - e^{-\Delta/kT})}{T_c(r)\log(1 - e^{-\Delta/kT_c})}\right)\,. \tag{5.49}$$

At sufficiently low temperature *all* particles are accommodated in the bound state at $-\Delta$, giving rise to a second paradox of ideal Bose-Einstein condensation: suppose the bound state wavefunction to be, qualitatively, $\Psi(\mathbf{r}) = C \exp -|\mathbf{r}|/\xi$, taking the position of the impurity to be the origin. $C = (2/\pi)^{1/2}/\xi$ is the constant of normalization and ξ the effective radius of the bound state. The single-particle density in this bound state is $\Psi^2(\mathbf{r})$; if N_Δ particles are bound, the particle density jumps to $N_\Delta\Psi^2(\mathbf{r})$.

We now estimate the *actual* particle density, $n(r)$. For $T \leq T_c$, $n(\mathbf{r}) = (1 - N_\Delta/N)N/L^2 + (N_\Delta/N)(2N/\pi)\xi^{-2} \exp{-2|\mathbf{r}|/\xi}$, i.e.

$$n(\mathbf{r}) = \left(\frac{T\log(1 - e^{-\Delta/kT})}{T_c\log(1 - e^{-\Delta/kT_c})} \right) \frac{N}{L^2}$$

$$+ \left(1 - \frac{T\log(1 - e^{-\Delta/kT})}{T_c\log(1 - e^{-\Delta/kT_c})} \right) \frac{2N}{\pi\xi^2} e^{-2|\mathbf{r}|\xi} . \tag{5.50}$$

In any regular homogeneous system, the particle density has to be intensive, i.e. independent of size in the thermodynamic limit. The first contribution to $n(r)$ in (5.50) is quite properly $O(N/L^2)$ but the term on the next line is quite different, as it is explicitly *extensive* over a finite-sized region of extent $O(\xi^2)$. And as such, it dominates the *rhs* of this equation at all temperatures $T < T_c$. This unphysical result comes about because an infinite number of particles, $N_\Delta = O(N)$, are "sucked into" a finite-sized region.

In essence the impurity potential has localized the Bose–Einstein condensation and caused it to occur in coördinate space! This phenomenon is not just restricted to low dimensionalities but can occur in 3D, whenever the particles are subject to attractive potential wells sufficiently deep or wide to have stable bound states.

None of this could happen if we included a hard core potential *ab initio*, one that allows only a finite number of particles to fit into a finite space. We next examine such the hard core repulsion in 1D, where the model can be solved in closed form.

5.11.   Consequences of a Hard Core Repulsion in 1D

We generalize the discussion of Tonks' gas in Sec. 4.9 and solve the quantum problem in closed form.

Start with a many-body wavefunction $\Phi(x_1, x_2, x_3, \ldots, x_N)$, all x in the range $0 < x < L$, subject to boundary conditions $\Phi = 0$ for any $|x_n - x_m| < b$, to reflect the presence of a hard core potential of diameter b. We assume a "natural" ordering, $x_n < x_{n+1}$. Although the initial ordering cannot change dynamically[1] we may construct different particle orderings by applying a *particle permutation* operator:

$$P\Phi(x_1, x_2, x_3, \ldots, x_N) = \Phi(x'_1, x'_2, x'_3, \ldots, x'_N), \quad \text{with } x'_1, x'_2, x'_3, \ldots, x'_N$$

---

[1]There can be no "tunneling" through a finite region of infinite, repulsive, potential.

a permutation of the coördinates $x_1, x_2, x_3, \ldots, x_N$. For bosons, symmetry dictates that $P\Phi(x_1, x_2, x_3, \ldots, x_N) = +\Phi(x_1, x_2, x_3, \ldots, x_N)$. Next, we solve the model by obtaining all the eigenstates, using the natural ordering $x_n < x_{n+1}$. Once this is achieved the preceding rules (symmetry under permutations of the particles) yield the states for arbitrary orderings of the particles.

First, change variables to $y_n = x_n - (n-1)b$. Now $\Phi(y_1, y_2, y_3, \ldots, y_N)$ satisfies a Schrödinger equation for free particles subject to the set of *boundary conditions*: $\Phi = 0$ if $y_1 = 0$, any $y_n = y_{n+1}$ or $y_N = L - (N-1)b$. When no two y's are equal, the problem is just that of N free particles between 2 walls (one at 0 and the other at $L^* \equiv L - (N-1)b$.) The eigenfunctions are: $\Psi(k_1, k_2, \ldots, k_N; y_1, y_2, \ldots, y_N) = \prod_n \Phi_{k(n)}(y_n)$ and constructed out of the normalized factors

$$\phi_k(y) = \sqrt{\frac{2}{L^*}} \sin ky \, , \qquad (5.51)$$

subject to $kL^* = m\pi$, with $m = 1, 2, 3, \ldots$ (This condition "quantizes" the k's in the product function.) There are $N!$ different Ψ's constructed by assigning the $k_{n's}$ to the $y_{m's}$. There exists just *one* linear combination of Ψ's that satisfies the given boundary conditions. It is:

$$\Phi(y_1, \ldots, y_N) = \frac{1}{\sqrt{N!}} \sum_P (-1)^P P\Psi(k_1, k_2, \ldots, k_N; y_1, \ldots y_N) \, .$$

Here the permutations P are not of the coördinates (denoted y_j) but of the set of k's; clearly this Φ is the *determinantal* function,

$$\Phi(y_1, \ldots, y_N) = \frac{1}{\sqrt{N!}} \det[\phi_{km}(y_n)] \, , \qquad (5.52)$$

by the very definition of a determinant. From this it follows that unless the k's are all distinct, two or more rows in the determinant are equal causing the determinant to vanish identically. According to the rules of quantum mechanics, a function that vanishes everywhere cannot be normalized and may not be used as an eigenfunction.

All the permutations of the k's have the same total energy, $E = \sum \frac{\hbar^2 k^2}{2M} n_k$. Because any given k is present at most once in a given determinant, the occupation-number operators n_k for each are restricted to the two eigenvalues, 0 and 1, just as for *fermions*. The ground state wave function, Φ_0, is *positive* everywhere.

We can express the energy of an arbitrary state by reference to the ground state energy E_0,

$$E_0 = \sum_{m=1}^{N} \frac{\hbar^2(\pi m)^2}{2ML^{*2}} \equiv N\frac{(\hbar\pi)^2}{6M}\left(\frac{N}{L^*}\right)^2\left[1 + \frac{3}{2N} + \frac{1}{2N^2}\right].$$

The highest occupied energy level ("Fermi level") is at $\varepsilon_F = \frac{\hbar^2(\pi N)^2}{2ML^{*2}}$. We best express these results by defining $1/r = N/L$ and $1/r^* \equiv N/L^* = 1/(r-b)$, i.e. $\varepsilon_F = \frac{\hbar^2\pi^2}{2Mr^{*2}}$ and $E_0 = (N/3)\varepsilon_F$. The following is, effectively, the Hamiltonian:

$$H = E_0 + \sum_{k>k_F} \frac{\hbar^2(k^2 - k_F^2)}{2M}\alpha_k^+\alpha_k + \sum_{0<k<k_F} \frac{\hbar^2(k_F^2 - k^2)}{2M}\beta_k^+\beta_k \qquad (5.53)$$

where α and β operate on states above and below the Fermi level, respectively. The eigenvalues of each occupation-number operator $\alpha^+\alpha$ and $\beta^+\beta$ is restricted to 0 and 1. At low temperatures all are 0 except for those within a range $\approx \pm 4kT$ from the Fermi level. For states within the narrow confines of this band of energy we can approximate the energy by 2 tangent curves at $\pm k_F$ setting $\frac{\hbar^2|(k^2-k_F^2)|}{2M} \approx \hbar|v_F||q|$. Here $v_F = \hbar\pi/Mr^*$ is the speed of particles at the Fermi level and $q = \pi|m - N|/L^*$ measures the "distance" from this reference level. For each of $\pm k_F$ the partition function of the α's (the levels above ε_F) is identical to that of the β's (those below ε_F) and therefore we need the 4th power of the one:

$$Z = e^{-\beta E_0} \times \left(\prod_q [1 + \exp -\beta\hbar v_F q]\right)^4.$$

The free energy is then

$$F = E_0 - 4kT\sum_q \log[1 + \exp -\beta\hbar v_F q],$$

i.e.:

$$F = E_0 - 4L(kT)^2 \frac{r^*/r}{\hbar\pi v_F}\int_0^\infty dy\, \log(1 + e^{-y})$$

$$= E_0 - 4L(kT)^2 \frac{r^*/r}{\hbar\pi v_F}\frac{\pi^2}{12}. \qquad (5.54)$$

Letting the hard core radius vanish, i.e. taking $\lim \cdot b \to 0$, does not eliminate $E_0 = N\hbar^2\pi^2/6Mr^{*2}$, which decreases only to the finite limiting value $N\hbar^2\pi^2/6Mr^2$. On the other hand, *without* the hard core the ground state energy *should* and *must* vanish as all particles would be in the lowest state.[m] But because the hard core affects boundary conditions ($\Phi = 0$ when two hard-core bosons collide,) it cannot be "turned off" merely by taking its radius to zero.

Owing to permutation symmetry, the Fourier transform of the ground state $\Phi(x_1, \ldots, x_N)$ undoubtedly exhibits an anomalously large $k = 0$ component — sufficiently for there to be at least a hint of "condensation". However, there is *not even a hint* of a critical T_c, although this is not unexpected in 1D. Note: if, instead of hard core bosons, we had considered hard core *fermions* in 1D, *the free energy would have remained precisely the same* as in (5.54) even though fermions require $P\Phi(x_1, x_2, \ldots, x_N) = (-1)^P \Phi(x_1, x_2, \ldots, x_N)$, the sign depending on whether the permutation is even or odd. *For in 1D a hard core preserves the particles' initial ordering, in which case the inherent "statistics" or permutational symmetry is irrelevant to the free energy.*

5.12.   Bosons in 3D Subject to Weak Two-Body Forces

The theory outlined below, originally proposed by Bogolubov[n] for weakly interacting bosons in 3D, was later generalized by Lee, Huang and Yang[o] to hard spheres at low density. The idea is simple: if bosons can be "kicked" out of the $k = 0$ state at finite T by thermal fluctuations, then so can they be ejected by the interparticle interactions. The bose condensation is then "nonideal", depending as it does on the nature and strength of the two-body forces. Let us start with the Hamiltonian of the nonideal system, consisting of kinetic and potential energies:

$$H = \frac{-\hbar^2}{2M} \sum_{j=1}^{N} \nabla_j^2 + g \sum_i \sum_{j \neq i} V(r_i - r_j). \qquad (5.55)$$

The potential energy has a short-range repulsive part and a longer range attractive part, and is possibly of the form $V(\mathbf{r}) = [f^2(|\mathbf{r}|) - 2f(|\mathbf{r}|)]$, with g measuring its strength and $f(|\mathbf{r}|)$, a monotonic decreasing function of $|\mathbf{r}|$ such as $\exp -(|\mathbf{r}| - r_m)/\xi$ or $|r_m/\mathbf{r}|^6$, determining its shape. A similar interaction potential was previously introduced and plotted in Sec. 4.6. When reëxpressed in *second quantization* (i.e. *using occupation-number operators*

---

[m] N particles would each have the minimum energy, $\hbar^2\pi^2/2ML^2$, so $E_0 \propto 1/rL \to 0$.
[n] N. N. Bogolubov, *J. Phys. USSR* **11**, 23 (1947).
[o] T. D. Lee, K. Huang and C. N. Yang, *Phys. Rev.* **106**, 1135 (1957).

attached to states labeled by k) the Hamiltonian becomes:

$$H = \sum_k \varepsilon(k) a_k^+ a_k + \frac{g}{2L^3} \sum_k \sum_{k'} \sum_q v(q) a_{k+q}^+ a_{k'-q}^+ a_{k'} a_k + \mu N , \qquad (5.56)$$

where $\varepsilon(k) = \frac{\hbar^2 k^2}{2M} - \mu$ and $gv(\mathbf{q}) = gv(-\mathbf{q})$, assumed real, is the *Fourier transform* of $gV(\mathbf{r})$. For ultimate simplification we have added and subtracted terms proportional to $\mu-$ just as in the grand ensemble. However, the Hamiltonian explicitly conserves particles in the following sense: if one starts with a state of N particles, matrix elements of (5.56) can only connect it to other states that also have precisely N particles. Therefore in solving for the eigenstates of H one should pick only those states $|\Psi)$ for which $\sum a_k^+ a_k |\Psi) = N|\Psi)$, so that both H and F are totally independent of μ.

Note that while number conservation was automatic (if implicit) in the "first quantized" representation of (5.55), in the modified, second quantized form of Eq. (5.56), it seems to have become somewhat of a chore! We rectify this henceforth by enforcing only the following, weaker, condition.

Let μ be fixed at a value that makes F stationary, i.e. $\partial F/\partial \mu = 0$, with N then allowed to vary about a specified mean value. The solution to $\partial F/\partial \mu = 0$ that optimizes F determines the "best" μ. The condition $\partial F/\partial \mu = 0$ is functionally equivalent to conserving the number of particles *on average*, i.e. to a constraint:

$$N = \sum_k \langle a_k^+ a_k \rangle_{TA} . \qquad (5.57)$$

We subject F and the eigenstates of H to this weaker condition only.

We now attempt to construct a complete set of solutions to Schrödinger's time- independent eigenvalue equation, $H|\Psi = E|\Psi)$, subject to Eq. (5.57). First we make the following shift, suggested by ideal bose condensation:

$$a_0^+ \to a_0^+ + \sqrt{n_0} \text{ and } a_0 \to a_0 + \sqrt{n_0}, \qquad (5.58)$$

in which n_0 is the macroscopic number that are expected to condense in the $k = 0$ mode at a given T and g. Once n_0 is properly chosen it will allow $\langle a_k^+ a_k \rangle$ to be small — at most $O(1)$ — for any k. After the shift (5.57) becomes:

$$N \to n_0 \sum_k \langle a_k^+ a_k \rangle + \sqrt{n_0} \langle a_0 + a_0^+ \rangle . \qquad (5.59)$$

The last term, $\langle a_0 + a_0^+ \rangle \equiv \langle q_0 \rangle$, actually vanishes as shown below.

H now contains several distinct parts. First, extensive constants:

$$H_0 = -\mu n_0 + \frac{gv(0)}{2L^3}n_0^2 + \mu N\,.$$

$(5.60a)$

Optimizing H_0 w.r. to n_0 yields $\mu = gv(0)n_0/L^3 \equiv gv(0)/r_0^3$, where $r_0 = L/n_0^{1/3}$ in 3D. This is a fortuitous choice, as it simultaneously serves to cancel two linear operators,

$$-\mu(a_0 + a_0^+)\sqrt{n_0} + \frac{2gv(0)}{2L^3}(a_0 + a_0^+)n_0^{3/2} = 0\,, \quad \text{from } H.$$

A *bilinear* contribution in the $k = 0$ field operators also appears in leading order in H, $H_0' = -\mu a_0^+ a_0 + \frac{gv(0)}{2r_0^3}[4a_0^+ a_0 + (a_0^2 + a_0^{+2})] = \frac{1}{2}gv(0)[(a_0 + a_0^+)^2 - 1]/r_0^3$. To this order the Hamiltonian does not contain operators of the form $(a_0 - a_0^+)$, therefore $q_0^2 \equiv (a_0 + a_0^+)^2$ is a constant of the motion. We set it equal to zero, as previouly promised, to minimize the energy. But, whatever its value, H_0' is not extensive and thus cannot contribute meaningfully to F.

The next terms are even less relevant to F, $\frac{gv(0)}{2L^3}[2(a_0^{+2}a_0 + a_0^+ a_0^2)\sqrt{n_0} + a_0^{+2}a_0^2]$, being $O(1/\sqrt{N})$ and $O(1/N)$ respectively. Thus the first nontrivial operator contributions are:

$$H_2 = \sum_{k\neq 0} \varepsilon(k)a_k^+ a_k$$

$$+ \frac{g}{2r_0^3}\sum_{k\neq 0}[v(k)(a_k a_{-k} + a_{-k}^+ a_k^+) + 2(v(0) + v(k))a_k^+ a_k]\,.$$

$(5.60b)$

The Hamiltonian H_2 yields the normal modes of the condensed system. It is easily reduced to quadrature[P] by combining terms in k and $-k$ and expressing it in the form,

$$H_2 = \sum_{k>0} h_k\,,$$

(5.61)

where $h_k = x_k(a_k^+ a_k + a_{-k}^+ a_{-k}) + y_k(a_k a_{-k} + a_{-k}^+ a_k^+)$ and

$$x_k = \frac{\hbar^2 k^2}{2M} - \mu + \frac{g}{r_0^3}(v(0) + v(k)) = \frac{\hbar^2 k^2}{2M} + y_k \quad \text{and} \quad y_k = gv(k)/r_0^3\,.$$

Each k-sector is diagonalized separately. After some calculation (to follow) one finds $h_k \rightarrow \tilde{h}_k = \omega_k(a_k^+ a_k + a_{-k}^+ a_{-k}) + \gamma_k$, exhibiting a phonon-like

---

[P]This quaint terminology derived from plane geometry means, more or less, "*eureka*".

spectrum at long wavelengths and a "binding energy" γ_k:

$$\left.\begin{array}{l} \omega_k = \sqrt{x_k^2 - y_k^2} = \sqrt{\dfrac{\hbar^2 k^2}{2M}\left(\dfrac{\hbar^2 k^2}{2M} + \dfrac{2gv(k)}{r_0^3}\right)} = \hbar k s(k)\,, \\[4mm] \text{with} \\[4mm] s(k) = \sqrt{\dfrac{gv(k)}{Mr_0^3}\left(1 + \dfrac{\hbar^2 k^2 r_0^3}{4Mgv(k)}\right)} \quad \text{and} \quad \gamma_k = \omega_k - x_k \end{array}\right\} \tag{5.62}$$

Also,

$$\langle a_k^+ a_k \rangle_{TA} = \frac{x_k}{2\omega_k}\coth\frac{1}{2}\beta\omega_k - \frac{1}{2}\,. \tag{5.63}$$

Summing up: after diagonalization the quadratic Hamiltonian contains only constants and number operators,

$$H_2 \to \sum_{k \neq 0} \omega_k \left(a_k^+ a_k + \frac{1}{2}\right) - \frac{1}{2}\sum_{k \neq 0} x_k \tag{5.64}$$

while the condition for (average) number conservation at finite T and g, Eqs. (5.57) and (5.59), takes on the appearance:

$$N = n_0 + \sum_{k \neq 0}\left(\frac{x_k}{2\omega_k}\coth\frac{1}{2}\beta\omega_k - \frac{1}{2}\right)\,. \tag{5.65}$$

The summand depends on n_0. As T is raised (at fixed g,) n_0 decreases monotonically from its value at $T = 0$, ultimately vanishing at the same T_c as in the ideal bose gas, Eq. (5.36). Above T_c the quadratic terms $\propto 1/r_0^3$ vanish altogether. (In 2D the integral in (5.65) is finite at $T = 0$ but diverges at any finite T, i.e. $T_c = 0$.)

To obtain these results the quadratic Hamiltonian was solved by means of Bogolubov's transformation. This conserves momentum but not particle number. Fortunately — in light of the discussion surrounding Eq. (5.57)*ff.* — the lack of number conservation is no longer a concern. The transformation mixes operators labeled by k with those labeled by $-k$, as follows:

$$\left.\begin{array}{l} a_k \to a_k \cosh\theta_k + a_{-k}^+ \sinh\theta_k \\[2mm] a_{-k} \to a_{-k}\cosh\theta_k + a_k^+ \sinh\theta_k \end{array}\right\} \tag{5.66}$$

The operators a^+ transform as the Hermitean conjugates of the above, e.g. $a_k^+ \to a_k^+ \cosh\theta_k + a_{-k}\sinh\theta_k$, etc. The parameter θ_k is a real, symmetric, function of k. It is trivial to verify that this transformation

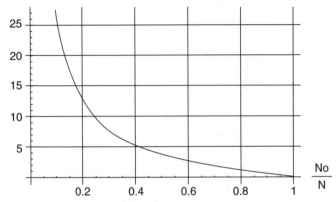

Fig. 5.4.   Interaction strength (in units of kT_c) versus n_0 at $T = 0$ in 3D, i.e. Ground State Depletion.
This plot, generated from Eq. (5.65) in lim $\cdot T = 0$, shows that the stronger the interaction the more depleted is the condensed phase.

- preserves Hermitean conjugation and
- preserves the commutation relations among the operators, (requirements that must be met in *any* unitary transformation.)

The transformed operators are inserted into the individual h_k and θ_k is chosen to eliminate off-diagonal operators such as $a_k a_{-k}$ and its Hermitean conjugate. This requires

$$\tanh 2\theta_k = -y_k/x_k \, . \tag{5.67}$$

After some more elementary algebra, the diagonal operator in Eq. (5.62) emerges.

The cubic and quartic terms are responsible for the scattering and breakup (finite lifetime) of the normal modes generated in H_2. Moreover they dominate the phase transition (the point at which the leading, quadratic, interactions vanish in proportionality to n_0). *Prior* to the Bogolubov transformation they were, respectively:

$$H_3 = \frac{g\sqrt{n_o}}{L^3} \sum_{k,q} v(q)(a_k^+ a_{k-q} a_q + a_q^+ a_{k-q}^+ a_k) \tag{5.68}$$

(all subscripts $\neq 0$ in the sums) and

$$H_4 = \frac{g}{2L^3} \sum_{k,k',q} v(q) a_{k'+q}^+ a_{k-q}^+ a_k a_{k'} \quad \text{(again, all subscripts} \neq 0\text{).} \tag{5.69}$$

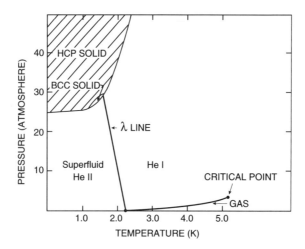

Fig. 5.5.   Low-T phase diagram for ^{4}He.

The transformation (5.66) replaces each factor a or a^+ by the sum of two operators, hence the summands in both (5.68) and (5.69) acquire 16 terms. At this point the theory loses its charm along with its simplicity,[q] although some of its features are in surprising accord with the remarkable properties of liquid helium. For example, at temperatures $T < T_c$, Eq. (5.62) yields a smooth spectrum $\omega(k)$ which is, at first, linear in k, rising smoothly to $\hbar^2 k^2/2M + $ const. The speed of sound $s(0)$ at small k can be fitted to experiment, as can "second sound", the entropy-temperature waves.

Still, Eq. (5.62) fails to yield the *roton* spectrum and the signature temperature-dependence of the roton gap, hallmarks of the neal superfluid ^{4}He to which we turn our attention next.

5.13.   Superfluid Helium (He II)

In the famous phase diagram of ^{4}He exhibited above, λ indicates the line of phase transitions separating normal He I from the superfluid He II. This last is one of the most interesting substances known. First liquefied by Kamerlingh Onnes in 1908, it is the founding member of the exclusive club of boson superfluids.

As the λ line is crossed, furious boiling ceases — as though the thermal conductivity or perhaps c_p, had suddenly diverged. The nature of the lambda

---

[q]However, the version of this theory adapted for *fermions* parallels the BCS theory which spectacularly explains and predicts thermodynamic properties of *superconductors*.

transition was subsequently discovered to be second-order and is associated
with with a logarithmically infinite heat capacity $C \propto \log 1/|1-T/T_\lambda|$. (The
shape of this curve, similar to the Greek letter, is what gives the transition
its name.)

Some aspects of the ideal bose gas and of its extension to the quadratic
Bogolubov theory are borne out in the superfluid phase. For example, the
condensed fraction n_o is not 1 at $T = 0$, but is reduced to $13.9 \pm 2.3\%$ by
the interactions. According to Fig. 5.4 this reduction requires an interaction
strength which, while quite weak, is approximately one order of magnitude
greater than kT_λ. This estimate is not too far from the mark. Given the light
mass of the He atoms the attractive portion of the van der Waals potential
is too shallow and short-ranged to allow a solid phase to be stable at $T = 0$
and atmospheric pressure. The lack of a solid phase at low p and T is borne
out in the phase diagram above.

The "$T^{3/2}$" law for the temperature-dependence of the decrease in n_o
derived in Eq. (5.39), should be replaced by a T^3 law when the dispersion
is linear, as in the Bogoliubov theory (rather than quadratic as in the ideal
bose gas.) This point is investigated in the following problem.

Problem 5.7. Using Eqs. (5.61)–(5.65), calculate and plot $1-n_0(T)/n_0(0)$
for several values of g in the Bogolubov theory, assuming in (5.62) $v(k) \approx$
$v(0)$, a constant. Derive the T^3 law analytically.

Experimentally[r] n_0 does follow a T^α law, with $\alpha = 3.6\pm1.4$, in qualitative
agreement with the Bogolubov theory. The low temperature heat capacity
also vanishes approximately as T^3 (again as expected for bosons with a linear
dispersion law, by analogy with the exact results for photons and phonons
derived earlier).

The dramatic decrease of viscosity in phase II, the decrease in inertia
of the superfluid, the quantization of circulation (rotational motion) and
the existence of novel branches of sound, including propagating waves of
oscillations in n_0 and T ("second sound"), all point to a rich spectrum of
elementary excitations. Explanations proposed by Landau and by Tisza involve 2 coëxisting fluids: one a normal liquid and the other a "condensed"
superfluid of n_0 particles devoid of viscosity, similar to what we had found
in the $k = 0$ state in the ideal bose gas.

[r]E. C: Svensson, *Proc. of Los Alamos Workshop* LA-10227-C, Vol. 2, 1984, p. 456.

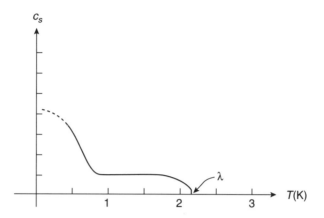

Fig. 5.6.   Speed of second sound c_s.
Theory predicts $c_s \to s(0)/\sqrt{3}$ ($s(0)$ = ordinary speed of sound) at $T = 0$.

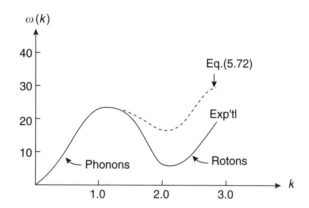

Fig. 5.7.   Experimental and theoretical spectrum of phonons and rotons.
Energy of elementary excitations $\omega(k)$ (in degrees K) versus wavenumber k (in Å^{-1})
Magnitude of the roton "gap" $D = \omega_{min}$ near $k = 2$ Å^{-1} is a function of T.

Feynman was able to deduce the spectrum in Fig. 5.7 out of a "variational" reconstruction of the elementary excitations.[s]

Let us suppose we knew the exact ground state Φ_0 and its energy
E_0 and use it to construct an operator $\Omega(k|r_1, p_1; \ldots; r_N, p_N)$ that carries
momentum k implicitly or explicitly. The variational excitation spectrum
is $\omega(k) = E(k) - E_0$, where $E(k)$ is the energy of the variational state

---

[s]In the variational method, an approximate solution is constructed and yields an upper
bound to the energy of the exact, but unknown state.

$\Omega(k|\cdots)\Phi_0$. Then,

$$\omega(k) = \frac{\langle\Phi_0|\Omega^+ H\Omega - \Omega^+\Omega H|\Phi_0\rangle}{\langle\Phi_0|\Omega^+\Omega|\Phi_0\rangle} = \frac{\langle\Phi_0|\Omega^+[H,\Omega]|\Phi_0\rangle}{\langle\Phi_0|\Omega^+\Omega|\Phi_0\rangle} \qquad (5.70)$$

The choice of Ω is crucial. Iff it is an exact raising operator of H the operator relation $[H,\Omega] = \omega'\Omega$ holds and (5.70) yield $\omega(k) = \omega'$.

The approximate raising operator introduced by Feynman in 1953 was simply $\Omega = \sum_{j=1}^{N} e^{ik\cdot r_j}$. With it, the numerator is,

$$-N\frac{\hbar^2}{2M}\langle\Phi_0|e^{-ik\cdot r_1}\sum_j[\nabla_j^2, e^{ik\cdot r_j}]|\Phi_0\rangle$$

$$= -N\frac{\hbar^2}{2M}\langle\Phi_0|e^{-ik\cdot r_1}\sum_j e^{ik\cdot r_j}\{2ik\cdot\nabla_j - k^2\}|\Phi_0\rangle$$

$$= N\hbar^2 k^2/2M . \qquad (5.71)$$

in which by permutation symmetry, one of the N identical particles is singled out, labeled "1" and its contribution multiplied by N.

Derivation: We have used the fact that Ω commutes with V which then drops out of the expession. Because Φ_0 is real, $\Phi_0\nabla\Phi_0 = 1/2\nabla(\Phi_0)^2$ appears in the integrand. A partial integration yields k^2, canceling $-k^2$ in the curly bracket *precisely* if $j \neq 1$. If $j = 1$ the partial integration yields 0. The final result is given in the last line.

The denominator in (5.70) is the ground state (or $T = 0$) value of $I(k) = NS(k)$ (cf. Eq. (4.26);) a plot of the experimental $S(k)$ at $T = 2$ K is shown in Fig. 4.4; it extrapolates to $S(k) \propto k$ at $T = 0$. Combining numerator and denominator one obtains,

$$\omega(k) = \hbar^2 k^2/(2MS(k)), \qquad (5.72)$$

an expression from theory that combines two independent measurements. The dashed curve in Fig. 5.7 reflects the accuracy of this formula. As subsequently refined by Feynman with the aid of his student M. Cohen, the variational fit to experiment has become even closer.[t]

[t]R. P. Feynman, *Phys. Rev.* **91**, 1291 and 1301 (1953), **94**, 262 (1954) and R. P. Feynman and M. Cohen, *Phys. Ev.* **102**, 1189 (1956).

The roton gap and minimum are revealed in the figure. The energy of an elementary excitation near the minimum is

$$\varepsilon(\mathbf{p}) \approx \Delta + \frac{(|\mathbf{p}| - p_{\min})^2}{2m^*} , \tag{5.73}$$

a function of $|\mathbf{p}|$ that is independent of angles. This form of dispersion allows the construction of stationary wavepackets of elementary excitations similar to smoke rings. Note that the velocity vanishes, $\nabla_p \varepsilon(p) = 0$, at the roton minimum.

A detailed discussion of the resulting vortex dynamics and of Feynman's speculations concerning the specific heat anomaly at the lambda point are to be found in the last chapter of his 1961 lectures on statistical mechanics.[u]

---

[u]R. P. Feynman, *Statistical Mechanics*, Addison-Wesley, Reading, 1998, p. 312.

Chapter 6

All About Fermions: Theories of Metals, Superconductors, Semiconductors

6.1. Fermi–Dirac Particles

Electrons, protons, neutrons, and a number of other distinct particles and "quasiparticles" have been found to satisfy the Pauli principle, which states that the wavefunctions $\Phi(1, 2, \ldots, N)$ of identical particles are totally anti-symmetric under any odd permutations, such as the simple interchange of any pair (i, j), and symmetric under any even permutation of the identical particles. The probability density $|\Phi(1, 2, \ldots, N)|^2$ thus remains symmetric under all permutations of the identical particles, as physically required. Insofar as we are primarily concerned with the statistical mechanics of a large number of particles, N will generally stand for a macroscopic, extensive quantity.

According to Dirac's relativistic quantum mechanics, not only does spin accompany Fermi–Dirac statistics but — with some exceptions — so does *charge*. The latter labels a "particle" or "antiparticle", i.e. the electron or the positron, two sides of the same coin. Because antiparticles play little if any role in statistical mechanics on earth,[a] wherever possible we shall avoid these issues in the present text. We saw in Sec. 5.11 that in 1D a hard-core interaction enforces a non-statistical "exclusion principle" on bosons, turning them into what are, effectively, *spinless* fermions. A similar transformation informs the spin operators of the Ising model in 2D. Therefore we shall consider fermions with, and without, a spin degree of freedom, treating spin, whenever it is relevant, as a label not much different from the band index attached to electron states in solids.

---

[a]Leaving statistical space physics including primordial matter, neutron stars, black holes, intergalactic dust, . . . , to specialized treatises.

Just as there were 2 kinds of bosons, some fermions are *conserved* and others *nonconserved*. The former can be studied either with ordinary quantum mechanics or with the aid of field theory (second quantization), whereas the latter (often called "quasiparticles") yield *only* to second-quantization. We encounter quasiparticles in connection with the Ising model, and also in metal physics and in superconductivity.

6.2. Slater Determinant: The Ground State

Because of the Pauli principle the eigenstates of noninteracting fermions bound to a common potential well cannot, despite the lack of correlations, be simple product states. However a linear combination of product states always "works". Let us show this for spinless fermions. Suppose the normalized one-particle eigenstates of the jth fermion of in this potential well to be $\varphi_k(r_j)$. The product state is $\Phi(r_1, \ldots, r_j, \ldots) = \Pi \varphi_k(r_j)$. Let P be any of the $N!$ permutation operators on the set of k's in this product. Construct the sum, $\Psi(r_1, \ldots, r_j, \ldots) = C \sum_P (-1)^P P \Phi(r_1, \ldots, r_j, \ldots)$. Each of the $N!$ product states in this sum is orthogonal to the others and each is normalized; therefore the over-all normalization factor is $C = 1/\sqrt{N!}$ Slater recognized that this totally antisymmetric function concides with his eponymous $N \times N$ determinant:

$$\Psi(\ldots, r_j, \ldots) = \frac{1}{\sqrt{N!}} \det[\varphi_{k_m}(r_j)] \qquad (6.1)$$

in which m labels the columns and j the rows, or vice versa. Clearly all the quantum labels k_m have to be distinct, as a secondary consequence of the Pauli principle (which originally only required that Ψ be antisymmetric in the $\mathbf{r}$'s). The implications are deep, as they ultimately allow the quantum numbers themselves to serve as identifiers of the indistinguishable fermions instead of the spatial coördinates.

The kinetic energy is the sum of the individual kinetic energies. Because the masses M of the indistinguishable particles are identical (isotopes are *not* indistinguishable!) the kinetic energy is symmetric (i.e. invariant) under permutations. Thus, each one of the $N!$ permutations in Ψ has an identical energy,

$$E(\ldots, \mathbf{k}_m, \ldots) = \sum_{m_x} \sum_{m_y} \sum_{m_z} \frac{\hbar^2 \mathbf{k}_m^2}{2M} \qquad (6.2)$$

with each $\mathbf{k}_m = (k_{x,m}, k_{y,m}, k_{z,m})$ a vector. The precise values of the k's depend on the boundary conditions. In Sec. 5.11 we used hard-wall boundary conditions in 1D but in 3D it proves more convenient to adopt *periodic* boundary conditions: $\varphi(x+L_x) = \varphi(x)$, and similarly for y and z. Assuming isotropy and $L_x = L_y = L_z = L$, this produces $\varphi_k(\mathbf{r}) = L^{-3/2}\exp i\mathbf{k}\cdot\mathbf{r}$ provided k is of the form $k_{x,m} = 2\pi m_x/L$, etc., i.e.

$$k_m = 2\pi(m_x, m_y, m_z)/L\,, \quad \text{where } m_x,\ m_y \text{ and } m_z \text{ are integers.} \quad (6.3)$$

The points m_x, m_y and m_z form a cubic grid of unit lattice parameter.

The ground state of N particles is special. Let us analyze it first. The occupied states have k within a "Fermi sphere" containing precisely N points. The radius k_F of this sphere is defined by,

$$N = \sum_{m_x}\sum_{m_y}\sum_{m_z} 1 \Rightarrow \frac{L^3}{(2\pi)^3}\frac{4\pi}{3}k_F^3\,. \quad (6.4)$$

k_F is intensive. To add (or subtract) a single particle it is necessary to find an unoccupied (or occupied) state close to the surface of this sphere, i.e. with energy close to $\hbar^2 k_F^2/2M$. This further identifies the chemical potential, $\mu = \hbar^2 k_F^2/2M$.

At finite temperature it is reasonable to expect that states within kT above the surface of the Fermi sphere will become partly occupied while those within kT below the Fermi surface will become partly depleted. For more precise information let us perform the statistical mechanics of the ideal Fermi–Dirac gas next.

6.3.   Ideal Fermi–Dirac Gas

The partition function Z_N is obtained by allowing each and every k a single choice: *occupied* (with an appropriate Boltzmann factor), or *unoccupied*. The counting is subject only to an over-all, global, requirement that the occupied states always add up to N precisely. Using a method introduced in Chapter 1 we write:

$$Z_N = \frac{1}{2\pi}\oint d\theta e^{iN\theta}\prod_k (1 + e^{-i\theta}e^{-\varepsilon(k)/kT})\,, \quad (6.5)$$

with $\varepsilon(k) = \hbar^2 k^2/2M$. After exponentiating the product the integral is evaluated using steepest descents. As before,

$$\frac{\partial}{\partial\theta}\left\{i\theta N + \sum_k \log(1 + e^{-i\theta}e^{-\varepsilon(k)/kT})\right\} = 0 \quad (6.6)$$

identifies the stationary point. It is recognized to be on the imaginary axis: $i\theta = -\beta\mu(\beta)$, with $\beta \equiv (kT)^{-1}$. Then, (6.6) yields:

$$N = \sum_k \frac{1}{e^{\beta(\varepsilon(k)-\mu(\beta))}+1} = \sum_k f(\varepsilon(k)) \qquad (6.7)$$

with $f(\varepsilon)$ the *Fermi function* sketched in the following figure.

To within some negligible corrections, the free energy $F = -kT\log Z_N$ is proportional to the logarithm of the maximum value of the integrand as determined in Eq. (6.6), i.e.

$$F = -kT\log Z_N = \mu N - kT\sum_k \log(1 + e^{-(\varepsilon(k)-\mu)/kT}). \qquad (6.8)$$

The *functional derivative* $-kT\delta/\delta\varepsilon\log F$ helps to obtain the same Fermi function $f(\varepsilon)$ of Fig. 6.1 by a more direct method. First, let us vary just one of the eigenvalues, $\varepsilon(k)$, $\varepsilon(k) \to \varepsilon(k) + \delta\varepsilon$. Then, $H \to H + \delta\varepsilon c_k^+ c_k$. By Eq. (4.4) the derivative of F w.r. to $\delta\varepsilon$ is $\langle c_k^+ c_k \rangle_{TA} \equiv f(\varepsilon(k))$,   *QED.*

The energy E is the sum of the individual energies. The *thermodynamic* definition $E = \partial(\beta F)/\partial\beta$ shows it to be so.

$$E = \frac{\partial}{\partial\beta}\left[\beta\mu N - \sum_k \log(1 + e^{-\beta(\varepsilon(k)-\mu)})\right] = \sum_k \varepsilon(k)f(\varepsilon(k)). \qquad (6.9)$$

In arriving at the above result one makes use of Eq. (6.7) to cancel out two contributions proportional to $\partial\mu/\partial\beta$. Equation (6.9) confirms $f(\varepsilon)$ as the *average occupancy* of a state of given energy ε. Although it is definitely *not*

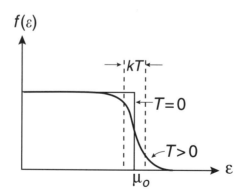

Fig. 6.1.   Fermi function versus energy at $T = 0$ and $T > 0$.
μ_0 is the chemical potential at $T = 0$. $\mu(T)$ deviates from μ_0 somewhat; see text for details.

a probability the "distribution function" $f(\varepsilon)$ is often mistaken for one, in that it is positive, ranges from 1 to 0, and that we average quantities over it.

Note that $f = 1/2$ at $\varepsilon = \mu$ precisely, at all T. Thus the locus of $f = 1/2$ determines the Fermi surface (spherical in our simple isotropic example). The volume of the Fermi sphere at $T = 0$ has to allow for N occupied one-particle states. This requirement uniquely determines the $T = 0$ value of the chemical potential μ_0.

The entropy is found most easily through $\mathscr{S} = (E - F)/T$, using the two equations above. Next we examine the effects of introducing an extra quantum number such as the spin.

6.4. Ideal Fermi–Dirac Gas with Spin

Under the ideal-gas assumption of no inter-particle interactions, Fermi–Dirac particles of spin "up" are statistically independent of those with spin "down", hence their partition functions factor. Assuming once again a total of N particles, $Z_N = Z_{N\uparrow}Z_{N\downarrow}$, with $N_\uparrow + N_\downarrow = N$. In the absence of an external magnetic field, $\mu_\uparrow = \mu_\downarrow = \mu$. Also, the spin polarization $M_z = 1/2(N_\uparrow - N_\downarrow)$, a quantity proportional to the total spin, vanishes.

Thus, $Z_N = (Z_{N/2})^2$, $F_N = 2F_{N/2}$, $E_N = 2E_{N/2}$, etc. But what is the energetic cost of establishing and maintaining a nonvanishing M_z?

$$F(N, M_z) = F_\uparrow + F_\downarrow = (\mu_\uparrow N_\uparrow + \mu_\downarrow N_\downarrow)$$
$$- kT \sum_k [\log(1 + e^{-(\varepsilon(k) - \mu_\uparrow)/kT}) + \log(1 + e^{-(\varepsilon(k) - \mu_\downarrow)/kT})].$$

$$(6.10)$$

When expanded about $M_z = 0$, to leading order in M_z this is expected to take the form $F(M_z) = F(0) + M_z^2/2\chi_0$ with $F(0) = 2F_{N/2}$ and χ_0 an extensive material property, the paramagnetic "spin susceptibility" of the Fermi-Dirac gas. It is both interesting and important to extract χ_0 from (6.10), for in the presence of an external magnetic field B (expressed in some appropriate units), $F = F(M_z) - BM_z$. When this F is minimized w.r. to $M_z (\partial F/\partial M_z = 0)$ it yields the usual low-field relation, $M_z = \chi_0 B$.

The sums in (6.10) as well as those in the preceding Eqs. (6.7) *ff.*, appear rather formidable *even in the limit* where they become integrals. Certainly the various quantities depend on dimensionality, as the density of states (*dos*) varies with dimensionality. Fortunately there exists a method for calculating Fermi integrals efficiently and accurately, in arbitrary dimensions, provided only that $\mu_\uparrow$ and $\mu_\downarrow$ do not differ much from each other and from μ and that $|M_z|$ is small (i.e. $\ll N$) and that $kT \ll \mu$.

6.5.   Fermi Integrals

Here we deal with the purely mathematical task of evaluating Fermi integrals of the type:

$$I(T,\mu) = \int d\varepsilon g(\varepsilon) f(\varepsilon). \tag{6.11}$$

with $g(\varepsilon)$ containing the *dos* and arbitrary other functions of ε.

Because the effects of finite T and B, the shifts in μ, etc. are all restricted to the vicinity of the $T = 0$ Fermi surface ("FS"), it is important to choose variables that reflect this. Write $f(\varepsilon) = \theta(\mu - \varepsilon) + \delta f$, with θ the usual step function,[b] μ the temperature-dependent $\mu(T)$, and δf expressed as follows in terms of a new dimensionless independent variable, $t = (\varepsilon - \mu)/kT$:

$$\delta f(t) = \frac{\text{sgn}(t)}{e^{|t|} + 1}, \quad \text{odd in } t \text{ and discontinuous at } FS. \tag{6.12}$$

Most functions $g(\varepsilon)$ that we shall encounter are smooth and featureless on a scale of kT, allowing for a Taylor series expansion,

$$g(\varepsilon) = g(\mu + tkT) = g(\mu) + t \left[\frac{kT}{1!} \frac{\partial g(\varepsilon)}{\partial \varepsilon} + t \frac{(kT)^2}{2!} \frac{\partial^2 g(\varepsilon)}{\partial \varepsilon^2} + \cdots \right]_{\varepsilon = \mu}$$

$$\equiv g(\mu) + t\Gamma(t, \mu). \tag{6.13}$$

Inserting the above into the integral one obtains a power series whose sole nonzero coefficients are *odd* derivatives of g (written as, $dg/d\varepsilon = g^{(1)}$, etc.) evaluated at the FS:

$$I(T,\mu) = I(0,\mu)$$

$$+ 2kT \sum_{n=0}^{\infty} (kT)^{2n+1} \zeta(2n+2)(1 - 2^{-(2n+1)}) g^{(2n+1)}(\varepsilon) \bigg|_{\varepsilon = \mu}. \tag{6.14}$$

Admittedly only an asymptotic expansion (due to neglect of terms $O(\exp -\mu/kT)$), this expansion converges rapidly to the exact results at

---

[b] $\theta(x) = 1$ for $x > 0$ and 0 otherwise.

low temperatures, $kT \ll \mu$. The derivation follows:

$$I(T, \mu) = \int_0^\mu d\varepsilon g(\varepsilon) + kT \left\{ \int_{-\mu/kT}^{+\infty} dt|t| \frac{\Gamma(t, \mu)}{e^{|t|} + 1} \right\}.$$

The error in extending the lower limit to $-\infty$ is just $O(\exp{-\mu/kT})$. (In typical metals the magnitude of μ/kT easily exceeds 40 at room temperature,[c] hence the estimated error is in the tenth decimal place!) In applications for which $g(\varepsilon)$ is a function only of ε and does not depend separately and explicitly on μ,

$$I(T, \mu) = \int_0^\mu d\varepsilon g(\varepsilon) + 2kT \left\{ \sum_{n=0}^\infty \frac{(kT)^{2n+1}}{(2n+1)!} \frac{d^{2n+1}g(\mu)}{d\mu^{2n+1}} \int_0^{+\infty} dt \, t^{2n+1} \frac{1}{e^{|t|} + 1} \right\}.$$

This last integral is evaluated in terms of the gamma and zeta functions: $\int_0^\infty dt \frac{t^{z-1}}{e^t + 1} = (1 - 2^{1-z})\Gamma(z)\zeta(z)$, so it only remains to set $z = 2n + 2$ to obtain the expansion in (6.14). To $O(kT/\mu)^6$, this expansion is explicitly:

$$I(T, \mu) = I(0, \mu) + (kT)^2 (\pi^2/6) \frac{dg(\mu)}{d\mu}$$

$$+ (kT)^4 (7\pi^4/360) \frac{d^3 g(\mu)}{d\mu^3} + \cdots \qquad (6.15)$$

with the first few terms sufficing for most practical purposes.

Problem 6.1.   Prove Eq. (6.12) and Eq. (6.13).

6.6.   Thermodynamic Functions of an Ideal Metal

Let us use this formula to evaluate the leading terms in the low temperature energy, heat capacity, entropy, paramagnetic spin susceptibility, and other thermodynamic properties of the ideal Fermi–Dirac gas in various dimensions. We start by expressing $E(T)$ as a function of T, N and μ and finish by obtaining $\mu(T)$ itself.

The calculation of E demonstrates the proper way to utilize the power expansion. In arbitrary dimensions d the *dos* of particles with quadratic dispersion ($\varepsilon \propto k^2$) is $\rho(\varepsilon) \propto \varepsilon^{d/2-1}$. Therefore in 3D the *dos* $\propto \sqrt{\varepsilon}$ and $E = N \int_0^\infty d\varepsilon \varepsilon^{3/2} f(\varepsilon) / \int_0^\infty d\varepsilon \varepsilon^{1/2} f(\varepsilon)$. Use of the ratio of Eqs. (6.7) and (6.9) allows the constants in the *dos* to cancel.

---

[c] $k_B \times$ room temperature is $\approx 1/40$ eV while μ lies in the range $1 - 10$ eV.

We carry out the expansion of numerator and denominator separately:

$E(T, \mu)$

$$= N \times \frac{\{(2/5)\mu^{5/2} + (kT)^2(\pi^2/6)(3/2)\mu^{1/2} - (kT)^4(7\pi^4/360)(3/8)\mu^{-3/2}\}}{\{(2/3)\mu^{3/2} + (kT)^2(\pi^2/6)(1/2)\mu^{-1/2} + (kT)^4(7\pi^4/360)(3/8)\mu^{-5/2}\}}$$

$$= \frac{3}{5}N\mu \frac{\{1 + (kT/\mu)^2(5\pi^2/8) - (kT/\mu)^4(7\pi^4/384)\}}{\{1 + (kT/\mu)^2(\pi^2/8) + (kT/\mu)^4(7\pi^4/640)\}}$$

$$= \frac{3}{5}N\mu\{1 + (kT/\mu)^2(\pi^2/2) - (kT/\mu)^4(\pi^4)(0.013541666\ldots)\}. \qquad (6.16)$$

Similarly, we express $N(\mu, T)$ in terms of $N_0 = N(\mu_0, 0)$:

$$N(\mu, T) = N_0 \int_0^\infty d\varepsilon\,\varepsilon^{1/2} f(\varepsilon) \Big/ \left(\frac{2}{3}\mu_0^{3/2}\right)$$

$$= N_0(\mu/\mu_0)^{3/2}\{1 + (kT/\mu)^2(\pi^2/8) + (kT/\mu)^4(7\pi^4/640) + \cdots\}. \qquad (6.17)$$

If N is held constant, Eq. (6.17) yields an implicit equation for $\mu(T)$ in terms of $\mu_0 \equiv \mu(0)$. If μ were constant (e.g. if the metal is connected to a battery) the equation yields the temperature-dependence of the particle number in the metal.

Assuming N = constant for an isolated metal, solve Eq. (6.17) to leading orders[d] in T/T_F after adopting the short-hand $kT_F \equiv \mu_0$:

$$\mu/\mu_0 = 1 - \frac{1}{12}(\pi T/T_F)^2 - \frac{1}{80}(\pi T/T_F)^4 - \frac{247}{25920}(\pi T/T_F)^6 + \cdots \qquad (6.18)$$

(Note: one could as easily have solved for $1/\mu$) At constant N:

$$E = \frac{3}{5}N\mu_0\left\{1 + \frac{5}{12}(\pi T/T_F)^2 - \frac{1}{16}(\pi T/T_F)^4 - \frac{1235}{36288}(\pi T/T_F)^6\right\}. \qquad (6.19)$$

One obtains the heat capacity by differentiating E, as given in Eq. (6.16) as a general function of $\mu(T)$ and T,

$$C_v(T) = \frac{\partial E}{\partial T}\Big|_v$$

$$= \frac{3}{5}N_0\{(d\mu/dt + \mu d/dT)(1 + (kT/\mu)^2(\pi^2/2) + \cdots)\}.$$

[d]E. Kiess, *Am. J. Phys.* **55**, 1006 (1987).

Whether at constant N or at constant μ, in leading order this quantity is linear in T, i.e. $C_v(T) = N_0\gamma T + O(T^3)$,

$$
\left.
\begin{aligned}
C_v(T) &= N_0 k_B(\pi^2/2)(T/T_F) + O(T/T_F)^3 \quad \text{(at const. } N)\\
&= N_0 k_B(3\pi^2/5)(T/T_F) + O(T/T_F)^3 \quad \text{(at const. } \mu)
\end{aligned}
\right\}
\quad (6.20)
$$

after returning the subscript B to Boltzmann's constant to emphasize that c_v has the units of k_B; its leading linear dependence on T is a generic property of metals[e] that is independent of dimensions.

Problem 6.2.   Typically one plots c_v/T versus T^2 to obtain the coefficients of the linear term (the intercept) and of the cubic term, the slope. (A) Evaluate all the corrections $O(T/T_F)^3$ in the expression (6.20) and (B) compare them, in magnitude, to the low-temperature T^3 Debye law derived in Sec. 5.6, assuming exactly one electron *per* cell (i.e. $N = N_0 = $ number of electrons = number of atomic cells). Identify the important parameters and determine whether it is possible the electronic T^3 contributions might obscure those of the phonons in an ordinary metal. Use the parameters: $T_F = O(10^6)$, $\theta_D = O(10^2)$.

Problem 6.3.   Rederive Eqs. (6.16)–(6.20) in closed form in 2D. Note the simplifications brought about by a constant *dos*. In what way do these results fail to be *rigorously exact*?

The equation of state of the Fermi–Dirac gas in 3D: $pV = 2E/3$, with E given in (6.19) is obtained most easily from a relation that was proved earlier for any and all ideal gases of particles with quadratic dispersion. Its entropy $\mathscr{S}$ is obtained in either of two ways: as an integral over C/T (see Sec. 3.4) or from $T\mathscr{S} = 5E/3 - \mu N$. The reader should verify that to leading order in T the quantities $\mathscr{S} = C$ are identical. (Therefore ordinary metals satisfy the Third Law.)

Problem 6.4.   Let $\mu_\uparrow = \mu + \Delta$, $\mu_\downarrow = \mu - \Delta$, in 3D. Calculate $M_z(\Delta, T)$ to leading orders in the parameters T/T_F and Δ/μ, both assumed small. Obtain the dependence of $\hat{\chi}_0(T) \equiv \lim \cdot \Delta \to 0 \{M_z(\Delta, T)/\Delta\}$ on T.

---

[e]And also of glasses and amorphous substances at low T, although for different reasons altogether.

6.7. Quasiparticles and Elementary Excitations

For many purposes it proves convenient to have as a starting point not
the absence of particles, but the ground state of the Fermi–Dirac gas. At
$T = 0$ all states below the FS are occupied and those above it are empty.
Deviations from this idealized state are described in terms of *quasiparticles*;
an electron added above the FS of a metal is a "quasielectron", while an
electron removed from below the FS of the same metal leaves behind a
"quasihole".

An "elementary excitation" of the Fermi sea is actual a compound,
created by promoting an electron at k of spin σ (σ is $\pm 1/2$, i.e. $\uparrow$ or $\downarrow$)
below the FS to k' and σ' above it. In the new terminology, this promo-
tion creates 2 quasiparticles: a positively charged quasihole at $-k$, $-\sigma$ and
a quasielectron at k', σ'. The signs are introduced to simplify conservation
of charge, momentum and spin angular momentum. In second quantization,
the transformation to quasiparticles consists of $c_{k,\sigma} \Leftrightarrow c^+_{-k,-\sigma}$ for $|k| < k_F$
and to no change above k_F. Then with $e_k = \hbar^2 k^2 / 2m - \mu$,

$$\sum_{k<k_F,\sigma} e_k c^+_{k,\sigma} c_{k,\sigma} \Rightarrow \sum_{k<k_F,\sigma} e_k (1 - c^+_{-k,-\sigma} c_{-k,-\sigma}).$$

On the *rhs* the sum without operators reproduces the ground state energy
E_0. Assuming inversion symmetry $e_{-k} = e_k$, the operator on the *rhs* becomes
$\sum_{k<k_F,\sigma} |e_k| c^+_{k,\sigma} c_{k,\sigma}$ and is *formally* the same as for $k > k_F$. Thus the total
excitation Hamiltonian, $H_0 - E_0$, is

$$H_0 - E_0 = \sum_{\text{all } k,\sigma} |e_k| c^+_{k,\sigma} c_{k,\sigma}. \tag{6.21}$$

The effective energy of a quasiparticle $\varepsilon_k \equiv |e_k|$ is plotted in the following
figure. Near the FS the symmetric cusp in quasiparticle energies reflects the
essential similarity of quasiholes and quasielectrons. This low-energy symmetry is quite general, as it is independent of dispersion or dimensionality.
(Any asymmetry in particles and holes, as evidenced in the figure, occurs
only at greater energies. It depends specifically on the dispersion but is significant only at high T.) The figure also illustrates the gap Δ introduced
into the quasiparticle spectrum in the BCS theory of superconductivity, to
be derived in Sec. 6.13.

The *dos* at the FS is $\rho(\mu) = 2\pi(2m)^{3/2}(2\pi\hbar)^{-3}\pi\sqrt{\mu(T)}$ and is assumed
constant within a few kT of $\mu(T)$. Under this assumption, certainly valid
at low T or in small external fields, an equal number of quasiholes and
quasielectrons are thermally produced. Thus the total number of physical

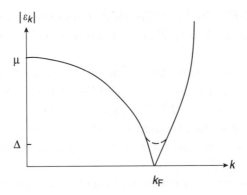

Fig. 6.2.   Quasiparticle Spectrum.
Normal metal, solid lines. Changes to the spectrum in the superconducting phase,
dashed segment.

particles is *automatically* conserved at finite T. The total internal energy is
merely,

$$E(T) = E_0 + L^3 4\rho(\mu) \int_0^\infty d\varepsilon \frac{\varepsilon}{e^{\varepsilon/kT} + 1}$$

$$= E_0 + L^3 4\rho(\mu)(\pi^2/12)(kT)^2 \qquad (6.22)$$

the factor 4 coming from the two factors: spin (factor 2) and quasiparticles (2 types). The Fermi integral was previously evaluated in Sec. 6.5; the
relevant zeta function $\zeta(2) = \pi^2/6$.

Using $N_0 = 2(4\pi k_F^3/3)L^3(2\pi)^{-3}$ to eliminate L^3 and setting $\mu \approx \mu_0 = k_B T_F$, one obtains $E(T) = E_0 + N_0 k_B (\pi/2)^2 T^2/T_F$ *in perfect agreement*
with the two leading terms in the expansion (6.19).

The corresponding specific heat is:

$$c_v(T) = 8\rho(\mu)(\pi^2/12)(k_B^2 T) = k_B(\pi^2/2)(T/T_F),$$

identical to leading order in T with a result previously obtained in Eq. (6.20)
by direct computation, assuming an isolated system at constant N.

Next we calculate the electronic paramagnetic susceptibility initiated in
Sec. 6.4 and sketched in Problem 6.4. At $T = 0$ "pour" a small number of
electrons from the surface of the spin-down Fermi sea onto the surface of the
spin-up Fermi sea. If $\mu_\downarrow$ is lowered by Δ the number of quasiparticle holes
involved is $L^3\rho(\mu)\Delta$ and the energy to create them is $L^3 1/2\,\rho(\mu)\Delta^2$. An equal
number of particles is then deposited onto the spin-up FS, raising its energy
also by $L^3 1/2\,\rho(\mu)\Delta^2$. Thus for a net magnetization $M_z = L^3\rho(\mu)\Delta$ the

cost in energy is $\Delta E = 2L^3 \times 1/2\,\rho(\mu)\Delta^2$. Comparison with $\Delta E = M_z^2/2\chi_0$ identifies $\chi_0 = L^3\rho(\mu)$. Finite T corrections are perforce $O(T/T_F)^2$ and are thus negligible.

Although $\rho(\mu)$ may not be known precisely, owing to uncertainties in m and μ, nevertheless in the case of an ideal Fermi-Dirac gas one predicts that the dimensionless ratio, $(C_v/k_B^2 T) \div (\chi_0)$, takes on a universal value $2\pi^2/3$ as $T \to 0$. (It should be emphasized that the derivation leading to this ratio is valid in any dimension and for any dispersion, as d and T influence only terms higher order in T.)

If however the particles are allowed to scatter and interact, as in fact they do in Nature, many-body effects *will* blur the FS and may change the ratio by a significant factor.

6.8.   Semiconductor Physics: Electrons and Holes

The generic dynamical, thermodynamical and electrical properties of insulators are, on their face, pretty dull. Fully occupied bands of "valence" electrons are separated by a substantial energy gap from unoccupied "conduction" bands. Any dynamics is mostly the result of atomic motion, i.e. *phonons*. One of the major discoveries of the twentieth century concerned the dilute doping of certain insulators by selected impurities, transforming them into *semiconductors*. To the extent that doping introduces electrons (*n*egatively charged particles) into the conduction bands they are called *n*-type; holes, i.e. positively charged carriers in the valence band, are the charge carriers in *p*-type semiconductors. The Fermi level generally lies *between* the bands where there are no allowed energy levels.

Photons of energy greater than the energy gap are absorbed in the material, where they release quasielectrons and quasiholes in equal numbers into the respective, separate, bands. These mobile carriers give rise to *photoconductivity*, a property useful in photography and in imaging. Figure 6.3 illustrates the energy spectrum in an *n*-type semiconductor and shows "donor" impurity states at an energy Δ below the conduction band edge, within the "forbidden" energy gap. The Fermi level μ lies, typically, at or below the donor level.

The energy of a conduction electron as measured from the band minimum is typically parabolic: $\varepsilon = p^2/2m_n^*$, with m_n^* a parameter of the band structure that is only loosely related to the free electron mass m_{el} and ranges from 10^{-2} to 10 times m_{el} in the various materials. It is measured in such experiments as *cyclotron resonance* that yield resonant microwave

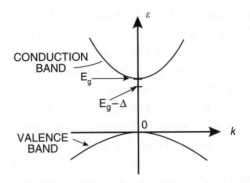

Fig. 6.3. Schematic band structure of a semiconductor.
This band structure is appropriate to an *n*-type "direct gap" semiconductor, in which the valence band maximum and conduction band minimum are both at $k = 0$. The donor levels at $E_g - \Delta$ are shown.

absorption at $\omega_c = eB/m_n^* c$. It can also be computed, more or less accurately, starting from first principles. The solid-state "band structure" calculations incorporate both the nature of the atoms and the geometrical structure, especially the periodicity, of the material.

Much the same holds for a hole, the energy of which is measured *down* from the the valence band maximum. The effective mass of a hole m_p^* is typically greater that that of the electron, m_n^*.

In what follows we shall specialize to *n*-type semiconductors. (Analogous results apply to *p*-types, *mutatis mutandis*.) The properties to be analyzed include the number of carriers N_{el} as a function of T, the number of neutral and ionized "donors", N_d^0 and N_d^+, the position of the chemical potential (Fermi level) as a function of T and last but not least, the statistical consequences of the electrons' Coulomb repulsion — sufficiently strong to prevent 2 electrons from being localized on a common donor.

6.9. *n*-Type Semiconductor Physics: The Statistics

Out of N atomic sites, a much smaller number N_d are substitutional impurities of valence +1 higher than the host, i.e. donor atoms. For hosts that are tetravalent such as germanium or silicon, the donors are typically pentavalent, e.g. phosporus. Because it is not covalently bonded the fifth electron is easily released ("donated") to the conduction band. Its binding energy Δ can be as small as 0.01 eV and, in any event, far smaller than $E_g \approx O(1 \text{ eV})$.

Thus at room temperature, $T \approx O(0.025 \text{ eV})/k_B$, the donors are pretty much ionized while the valence band remains full occupied.

The excess electrons have just two places to go: either into a bound state on one of the available donors or as free electrons in the conduction band. At first we shall treat the spin degeneracy in a trivial way, multiplying all the extensive quantities by a factor 2, postponing discussion of the Coulomb interaction to the following section. The maximum number of carriers in the conduction band is small, at most $N_d \ll N$, well within the dilute limit where the Boltzmann approximation to the Fermi–Dirac function is known to become valid. With N_d^0 being the number of occupied donor levels, and the number of ionized donors being $N_d^+ = N_{el}$, the equation that regulates the donors is $N_d = N_d^0 + N_d^+$, i.e.

$$N_d = 2\frac{N_d}{e^{\beta(-\Delta-\mu)}+1} + 2\left(\frac{L}{2\pi\hbar}\right)^3 \int d^3p \frac{1}{e^{\beta(\varepsilon(p)-\mu)}+1}$$

$$\approx 2\frac{N_d}{e^{\beta(-\Delta-\mu)}+1} + 2\left(\frac{L}{2\pi\hbar}\right)^3 \int d^3p\, e^{-\beta(\varepsilon(p)-\mu)} \qquad (6.23)$$

Evaluating the integral on the second line yields

$$N_{el} = 2\left(\frac{L\sqrt{2\pi m_n^* kT}}{2\pi\hbar}\right)^3 e^{\beta\mu}.$$

Thus reduced to a quadratic in $\exp(\beta\mu) \gg 1$, Eq. (6.23) is readily solved for the relevant quantity, $\mu(T)$. Next we discuss this equation in greater detail.

6.10.   Correlations and the Coulomb Repulsion

According to the preceding section, one might think that the *a priori* probability that any given donor is neutral (e.g. carries $4+1 = 5$ valence electrons) and has spin $+1/2$, i.e. that the fifth electron has spin "up", is just given by the Fermi function, $f = \frac{1}{e^{\beta(-\Delta-\mu)}+1}$. If so, the inclusion of both spin orientations should yield twice that, $2f = \frac{2}{e^{\beta(-\Delta-\mu)}+1}$. At $T = 0$, $\mu \to -\Delta$ (cf. lower curve in Fig. 6.4), and $f \to 1/2$; thus there are $N_d/2$ electrons with spin "up" and $N_d/2$ with spin "down" in the ground state of the donors, given that the respective probabilities are equal.

But this gives rise to a paradox. Insofar as the two species of electrons are uncorrelated some donors are then, statistically, occupied by two electrons of opposite spin (and thereby transformed into *negative ions*). The number of such ions is $f^2 N_d = (\frac{1}{e^{\beta(-\Delta-\mu)}+1})^2 N_d$, a quantity that approaches $N_d/4$ as

$T \to 0$. Clearly this result is implausible, given that the sum $(2f + f^2)$ will exceed 1 at low T! To evaluate the consequences of statistical correlations and to understand the important rôle played by Coulomb interactions in keeping an ion neutral, one has to return to "first principles": to the partition function and to a definition of the probabilities.

Consider the contribution to the partition function of a single donor site that can sustain $n = 4, 5$, or 6 electrons *in the absence* of the Coulomb interaction:

$$Z_d = 1 + 2e^{-\beta(-\Delta-\mu)} + e^{-2\beta(-\Delta-\mu)} \tag{6.24}$$

where the Boltzmann exponents are $-(n-4)\beta(-\Delta-\mu)$ respectively. The two occupation numbers $n = 4$ electrons and $n = 6$ electrons are unique, carry total spin 0, and correspond to the positive and negative ions respectively. The occupation $n = 5$ corresponding to a *neutral* donor carries the choice of spin "up" or "down", hence the extra factor 2. When normalized by Z_d each of the individual terms in the sum *is* a genuine probability as exhibited below.

Thus, at arbitrary T the *probability* of the donor being neutral is *not* $2f$ but rather,

$$P_0 = \frac{2e^{-\beta(-\Delta-\mu)}}{1 + 2e^{-\beta(-\Delta-\mu)} + e^{-2\beta(-\Delta-\mu)}} = \frac{1}{2}\text{sech}^2\beta(\Delta + \mu).$$

On the other hand the probability of a negative ion occupied by one electron of spin "up" and one of spin "down", is

$$P_- = \frac{e^{-2\beta(-\Delta-\mu)}}{1 + 2e^{-\beta(-\Delta-\mu)} + e^{-2\beta(-\Delta-\mu)}} = \frac{1}{(1 + e^{\beta(-\Delta-\mu)})^2}.$$

Finally the probability of a positive ion, unoccupied by electrons of either spin, is just

$$P_+ = \frac{1}{1 + 2e^{-\beta(-\Delta-\mu)} + e^{-2\beta(-\Delta-\mu)}} = \frac{1}{(1 + e^{-\beta(-\Delta-\mu)})^2}.$$

By construction these probabilities do add up exactly to 1 at all T.

The Fermi function reflects *average occupancy* by an electron of either spin and its resemblance to a probability is purely coïncidental. In the present case f is,

$$f(-\Delta) = \langle c_{d,\uparrow}^* c_{d,\uparrow} \rangle = \frac{0 + e^{-\beta(-\Delta-\mu)} + e^{-2\beta(-\Delta-\mu)}}{Z_d}$$

$$= \frac{e^{-\beta(-\Delta-\mu)} \times (1 + e^{-\beta(-\Delta-\mu)})}{(1 + e^{-\beta(-\Delta-\mu)})^2} = \frac{1}{e^{\beta(-\Delta-\mu)} + 1}$$

and when multiplied by 2 (for spin orientation) $\times N_d$ (for number of available sites) it yields the first term on the *rhs* of Eq. (6.23), but it is clearly *not* the correct value, $P_0 N_d$.

In the lim $\cdot T \to 0$, $\mu \to -\Delta$ and each of the exponentials in $Z_d \to 1$ and $Z_d \to 4$. Thus at the Absolute zero the probability of a negative or positive ion $\to 1/4$ each, and of a neutral ion $\to 1/2$. This is how we arrived at the conclusion that at $T \to 0$ the total number of bound electrons equals the number of donor sites. But the existence of negative ions on some sites offends the physics. Consider the following facts:

The lowest energy on a given donor site is that of the neutral configuration; the energy of a positive ion exceeds it by Δ at the least, and therefore it *should not be present* at $T = 0$. However the negative ion has a much higher energy $U \gg \Delta$ and is entitled to *even less* of a presence. Physically, the quantity U comes from the additional Coulomb repulsion of the excess electron orbiting a common donor center. Thus the correct Z_d is not (6.24), but rather, $Z_d = 1 + 2e^{-\beta(-\Delta-\mu)} + e^{-2\beta(-\Delta-\mu)}e^{-\beta U}$. In lim $\cdot U \gg kT$ the last term drops out altogether, and

$$Z_d \Rightarrow 1 + 2e^{-\beta(-\Delta-\mu)} . \tag{6.25}$$

With the aid of this function, the *correlated* probabilities are found to be:

$$P_0 = \frac{2e^{-\beta(-\Delta-\mu)}}{1 + 2e^{-\beta(-\Delta-\mu)}} = \frac{2}{e^{\beta(-\Delta-\mu)} + 2} , \qquad P_+ = \frac{1}{1 + 2e^{-\beta(-\Delta-\mu)}} ,$$

and $P_- = 0$, quite a different microscopic result! With Eq. (6.23) now discredited, the *correct* equation $N_d = N_d^0 + N_d^+$ has to be,

$$N_d = \frac{2N_d}{e^{\beta(-\Delta-\mu)} + 2} + 2\left(\frac{L}{2\pi\hbar}\right)^3 \int d^3p \frac{1}{e^{\beta(\varepsilon(p)-\mu)} + 1}$$

$$\approx \frac{2N_d}{e^{\beta(-\Delta-\mu)} + 2} + 2\left(\frac{L}{2\pi\hbar}\right)^3 \int d^3p\, e^{-\beta(\varepsilon(p)-\mu)} . \tag{6.26}$$

Because the "free" electrons in the conduction band are in extended states one is justified in neglecting the Coulomb interaction among them. Equation (6.26) bears a superficial resemblance to (6.23) but its solution is quite different in the details, with μ starting at $-\Delta/2$ at $T = 0$ and decreasing from there. The correct and the incorrect solutions, with and without correlations, are compared in the figure that follows.

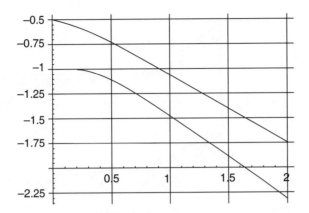

Fig. 6.4. $\mu(T)/\Delta$ versus kT/Δ with (better theory, upper curve) and without (less physical) Coulomb Repulsion on Individual Donor sites.

Here $\mu < 0$ as measured from the bottom of the upper (conduction) band.

6.11. Miscellaneous Properties of Semiconductors

1. Density of Carriers. Previously we found that the density of free electrons in the conduction band was, to good accuracy,

$$N_{el}/L^3 \equiv n = 2(2\pi\hbar)^{-3}(2\pi m_n^* kT)^{3/2}e^{\beta\mu} \,.$$

By analogy,

$$N_h/L^3 = p = 2(2\pi\hbar)^{-3}(2\pi m_p^* kT)^{3/2}e^{-\beta(\mu+E_g)}$$

is the density of holes in the valence band. It follows that the product $n \times p \equiv n_i^2$ is independent of μ and therefore of the density of donors or acceptors. Define $n_i \equiv 2(2\pi\hbar)^{-3}(2\pi kT)^{3/2}(m_n^* m_p^*)^{3/4}e^{-\beta E_g/2}$, the so-called "intrinsic" carrier concentration, a function only of the m^*'s, E_g and T. In an undoped semiconductors (no donors or acceptors,) $n = p = n_i$ for reasons of electrical neutrality. In that case the Fermi level lies approximately mid-gap, $\mu(T) = -E_g/2 + (3kT/4)\log(m_p^*/m_n^*)$.

2. Rectification. If a semiconductor is abutted to a metal a net flow of charge from one material to the other is required to equalize the μ's. The excess charge at the interface forms a dipole layer known as a "Schottky barrier", and is responsible for the rectification. The "cat's whiskers" germanium diodes used in radio reception in the 1920's were elevated to the status of an amplifying triode, the *transistor* as it was named by its inventors Bardeen, Brattain and Schockley in the late 1940's. Figure 6.5, based on a drawing in John Bardeen's 1962 Nobel address, shows how the electro-

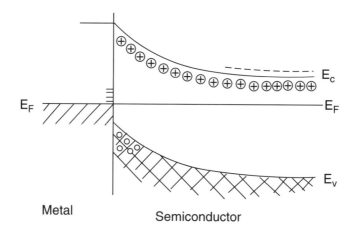

Fig. 6.5.   Energy level scheme of metal-semiconductor diode.
The semiconductor that is n-type in the bulk ($x \to +\infty$; conduction-band electrons
indicated by "$- - -$") becomes p-type in the vicinity of the metal surface (holes
symbolized by "$\circ \circ \circ \ldots$" in the valence band). In thermodynamic equilibrium the
electrostatic dipole layer lifts the semiconductor's energy levels at the surface relative to those in the bulk, in an amount approaching $E_g = E_c - E_v = O(1 \text{ eV})$, the
height of the *Schottky barrier*, while maintaining $\mu = E_F$ constant (so as to ensure
no current flow). Surface energy-states are indicated as short horizontal lines.

static forces "bend" the bands while keeping the Fermi level constant in
equilibrium.
3. Junctions. A similar barrier created at the interface of two *dissimilar*
semiconductors, e.g. of an n-type and a p-type, is known as a p-n junction.
For details refer to texts specialized in semiconductors. The following
illustrates a *Schottky diode*.

In thermal equilibrium a small current I_+ flows from left to right, compensated exactly by I_- flowing in the opposite direction. Let us call these quantities I_0. Now let us suppose that a potential V is applied to the barrier and
is dropped almost entirely in the high-resistance (semiconductor) vicinity of
the barrier, i.e. $\mu = E_F + V$ on the semiconductor side. The net current in
the diode is then

$$I(V) = I_0 e^{\beta eV} - I_0 , \qquad (6.27)$$

limited to $-I_0$ if $\beta eV < 0$ and exponentially large if $\beta eV > 0$. At small
$|\beta eV|$, i.e. for voltages $< 1/40$ V at room temperature, the current is ohmic.
In a triode a *third* electrode capable of changing $V \to V + \delta V$ permits an

exponentially larger change in current at a given δV, hence produces signal- and possibly power-amplification.

6.12. Aspects of Superconductivity: Cooper Pairs

The BCS theory of superconductivity,[f] based on the pairing of electrons of opposite spins and momenta, was spawned in a brief interval in the years 1956–1958. Prior to that, a number of competing theories sought to explain aspects of this mysterious phenomenon discovered in 1911; a crude time-line follows.[g]

By 1934, Gorter and Casimir's theory of super conductivity as a two-fluid model predicted a thermodynamic second-order phase transition and by 1935, F. and H. London's theory of magnetic flux exclusion indicated a "stiff" ground state wavefunction. By 1950 the equations of Ginzburg and Landau allowed calculation of inhomogeneous properties, especially near surfaces. Already an "isotope effect" had been observed, implicating phonons in what was otherwise a purely electrical phenomenon. Most significantly, all experiments were compatible with the existence of a temperature-dependent *electronic energy gap* pinned — by a force or forces yet to be identified — to the Fermi surface of each superconductor. The difficulties lay primarily in the fomulations, and not in the lack of physical understanding. With some exceptions (superfluidity, lattice vibrations, ferromagnetism), prior the BCS theory in 1956 many-body theory had yet to be successfully formulated and applied to realistic problems in solid-state physics.

Superconductivity was *not* a ubiquitous phenomenon. It seemed to involve a small fraction of the charge carriers, possibly fewer than 1 out of 10,000. In metals such as mercury (Hg), lead (Pb) or niobium (Nb) or their alloys the room-temperature resistivity is an order of magnitude higher than in copper (Cu), the reference material. Yet it is just these "poor" metals, not copper, that superconduct at characteristically low temperatures, $T < T_c \approx 10$ K.

In a 1956 address, Feynman noted[h] that characteristic energies of normal metals are approximately 10,000–100,000 K while those of a superconductor are 10 K. (In this one respect superconductivity differs markedly from

---

[f]See references in Chapter 3, footnote 19.

[g]These historical references are from the review "Recent developments in superconductivity" by J. Bardeen and J. R. Schrieffer, Chapter 6 in Vol. III, *Progr. Low Temp. Phys.*, C. J. Gorter, Ed., North-Holland, 1961.

[h]R. P. Feynman, *Rev. Mod. Phys.* **29**, 205–212 (1957).

superfluidity, in which the characteristic phenomena have energies of the same order of magnitude as the potentials.) Considering that the principal one-particle energies: the band-stucture, the Fermi energy, the two-body interactions, etc., are never known to better than a few %, we might expect the search for a quantitative theory of superconductivity to be like a quixotic search for one particular grain of sand at the beach. In conclusion, Feynman counseled that

> "...what we must do is not compute anything, but simply guess what makes the ground state isolated at a lower energy. That is, guess what kind of correlation exists at long distances. Why haven't we theoretical physicists solved this problem yet? ...It has nothing to do with experiments.... The only reason that we cannot do this problem of superconductivity is that we haven't got enough imagination."[h]

But there *was* an imaginative breakthrough and it came just a few months later.

Trained as a "particle" physicist, Leon Cooper was a post-doctoral assistant to John Bardeen at the University of Illinois in 1956 when he conjured up the following model of 2 electrons bound in a singlet configuration, with total momentum $\hbar q$. This "Cooper pair" was constrained to live outside the Fermi sphere, but was bound by attractive forces generated through the exchange of a virtual phonon. At $q = 0$ the pair's wavefunction is,

$$\Psi(1,2) = \sum_k \phi_k \frac{\cos k \cdot (r_1 - r_2)}{\Omega} (\uparrow_1\downarrow_2 - \downarrow_1\uparrow_2) \,,$$

where ϕ_k satisfies a Schrödinger equation:

$$(2e_k - E)\phi_k - \frac{U}{\Omega} \sum_{k'} \phi_{k'} = 0 \,. \tag{6.28}$$

Here, $e_k = \frac{\hbar^2 k^2}{2m} - \mu$, Ω is the volume and $E < 0$ is the energy of the bound pair. All k are restricted to the range $0 < e_k < k_B\theta_D$. By definition all states of energy $e < 0$ are occupied and therefore unavailable. The upper limit on e_k is the maximum energy of a phonon, a natural cut-off inherent in the phonon-exchange mechanism.[i] To solve the above equation, replace

---

[i]This last seemed the most plausible mechanism for the existence of a short-ranged, velocity-dependent, attractive potential $-U$ in metals, and was in accord with the isotope effect. Various other mechanisms had been proposed and debated. Indeed, a distinctly different mechanism is presumed responsible for the copper oxides exhibiting *high-temperature* superconductivity at $O(100 \text{ K})$.

the sum by an integration over $\rho(\mu)$, the one-particle *dos* evaluated at the Fermi surface.

Thus, for $E < 0$, $\phi_k = \phi(e) = U\rho(\mu)\Gamma(2e + |E|)^{-1}$, where $\Gamma \equiv \sum_{k'} \phi_{k'}$ is the "lumped" parameter,

$$\Gamma = \{U\rho(\mu)\Gamma\} \left\{ \int_0^{k\theta_D} de \frac{1}{2e' + |E|} \right\} = \left\{ \frac{1}{2} U\rho(\mu)\Gamma \right\} \{\log(1 + 2k\theta_D/|E|)\} .$$

This equation always has a trivial solution $\Gamma = 0$. The nontrivial is,

$$E = -\frac{2k\theta_D}{2^{2/U\rho(\mu)} - 1} \approx -2k\theta_D e^{-2/U\rho(\mu)} . \tag{6.29}$$

Above the bound state there lies a continuum of scattering solutions to Eq. (6.28), of energies $2e_k + O(1/\Omega) > 0$, separated from the bound state by an "energy gap" $|E|$.

By 1956 the "*isotope effect*" had already been established in some super-conducting metals for which numerous isotopes are readily available, such as Hg, Sn, Pb, Th. Collective properties of the electron gas, including the energy gap and T_c, together with the "critical" magnetic field $B_c(0)$, *all* scaled with the atomic (*isotopic*) mass $\sqrt{1/M}$, just as does θ_D.[j] Cooper's formula, Eq. (6.29), linking electronic properties to the lattice vibrations, yields a natural[k] explanation of the dependence of a sensitive electronic property on atomic mass. Additionally, the very *existence* of an energy gap explains the exponentially small specific heat of superconductors observed at low temperatures $T \ll T_c$, as shown in Fig. 3.9. The seeming paradox that the "worst" metals (those with the highest room-temperature electrical resistance) make the "best" superconductors (highest T_c), received a plausible explanation in this type model: an increase in electron-phonon scattering increases the effective attraction U, hence boosts $|E|$ and T_c at the same time it decreases the electrons' mean-free path.

Despite numerous inconclusive details the excellent results obtained from such a "toy" model were tantalizing to Bardeen, Cooper and — most especially — to Bardeen's graduate student Bob Schrieffer, who had undertaken the theory of superconductivity as his Ph.D. project.[l] Schrieffer soon "guessed" the imaginative many-body gound state satisfying all the requirements that Bardeen and Cooper had set out in earlier,

---

[j]See §5.6, Eq. (5.28a) and the discussion on Θ_D, s, etc.
[k]Provided U is independent of M, as it generally is found to be.
[l]For an historical overview of the original BCS theory, see J. R. Schrieffer, *Theory of Superconductivity*, Benjamin, New York, 1964.

semiphenomenological, analyses. All three were to share the Nobel prize for
their efforts, an unprecedented second such honor for John Bardeen.

6.13.   Aspects of BCS Theory

In its simplest form, this many-body theory of superconductivity pairs each
electron of momentum $\hbar k$ and spin "up" to an electron of opposite momentum and spin. The result is a collective state with no net magnetic moment
that carries no current. Elementary excitations are separated from the gound
state by an energy gap that is temperature dependent and vanishes at T_c. As
we shall see, the very *form* of the solution suffices to explain both the complete lack of resistance and the *Meissner effect* (as the expulsion of magnetic
field from the bulk of the superconductor is called).

It proves useful to introduce the fermion operators in occupation-number
space: creation and annihilation operators c^+ and c. Unlike boson operators,
these *anticommute*: $c_{k,\sigma}c_{k',\sigma'} + c_{k',\sigma'}c_{k,\sigma} = 0$. The anticommutator or "curly"
bracket is just $\{A, B\} \equiv AB + BA$, i.e.

$$\{c_{k,\sigma}, c_{k',\sigma'}\} = 0 = \{c^+_{k,\sigma}, c^+_{k',\sigma'}\}, \quad \text{and} \quad \{c_{k,\sigma}, c^+_{k',\sigma'}\} = \delta_{k,k'}\delta_{\sigma,\sigma'} \quad (6.30)$$

k labeling the wave-vector of a free electron and $\sigma = \downarrow$ or $\uparrow$ (i.e. $\pm 1/2$) its
spin. Differently labeled pairs of such operators *commute*:

$$[c_{k,\sigma}c_{-k,-\sigma}, c_{k',\sigma'}c_{-k',-\sigma'}] = 0\,.$$

In this sense, paired fermion operators resemble the boson operators of
Eq. (5.12). Ultimately this proves to be a false cognate, as paired fermions
fail to satisfy the totality of commutator relations required of bosons. In
particular, $[c_{k,\sigma}c_{-k,-\sigma}, c^+_{-k,-\sigma}c^+_{k,\sigma}]$ is an operator that is $\neq 1$ or 0. This is the
principal reason paired fermions are not generally considered to be bosons
unless the binding energy is enormous.

The selection of *which* Hamiltonian to diagonalize hinges on the total
current carried by the system. (In the absence of persistent currents and
of magnetic fields the total momentum and total spin are both zero.) At
arbitrary current densities the relevant interactions included by BCS in their
famous "reduced" Hamiltonian consists of kinetic energy, plus that part of
the attractive interactions compatible with momentum conservation:

$$H_{\text{eff}}(q) = \sum_k (e_{k+q/2}c^+_{k+q/2,\uparrow}c_{k+q/2,\uparrow} + e_{k-q/2}c^+_{-k+q/2,\downarrow}c_{-k+q/2,\downarrow})$$

$$- \frac{U}{\Omega}\sum_{k,k'} c^+_{k+q/2,\uparrow}c^+_{-k+q/2,\downarrow}c_{-k'+q/2,\downarrow}c_{k'+q/2,\uparrow}\,. \quad (6.31)$$

The rest of the Hamiltonian is treated by perturbation theory and is found to be largely irrelevant. At $\mathbf{q} = 0$ the variational solution,

$$|\Psi\rangle \equiv \sum_{k\theta_D > e_k > 0} \phi_k c_{k,\uparrow}^+ c_{-k,\downarrow}^+ |F\rangle \,,$$

with $|F\rangle$ the filled Fermi sphere, satisfies Cooper's Eq. (6.28) and yields his ground state energy in (6.29). But note that this E was only $O(1)$ lower than the unperturbed energy of the Fermi sphere, whereas *the ground state energy of this many-body problem has to be extensive!*

BCS proposed a current-carrying ground state of the type,[m]

$$\prod_{\text{all } k} \{\sin \phi_k + \cos \phi_k c_{k+q/2,\uparrow}^+ c_{-k+q/2,\downarrow}^+\}|0\rangle \tag{6.32}$$

with $\mathbf{J} = eN_{el}\hbar q/m$ the current, and ϕ_k a function to be determined variationally for the lowest ground state energy. The model's dynamical and thermodynamical properties were derived using an ingeniously constructed complete set of orthonormal states.[l]

Alternatively, the ground state can be found by linearizing the Hamiltonian. This procedure works only because quartic terms in the interaction can be factored into the product of two macroscopic quadratic sums. For example, $\Delta_{op}^* \equiv \frac{U}{\Omega} \sum_k c_{k,\uparrow}^+ c_{-k,\downarrow}^+$ is an intensive quantity $O(1)$ whose fluctuations have to be negligible, $O(1/\sqrt{\Omega})$. After replacing Δ_{op}^* by its average Δ^* and adding a trivial constant to the Hamiltonian (6.31), it is linearized as follows (at $\mathbf{q} = 0$):

$$H = \sum_k e_k(c_{k,\uparrow}^+ c_{k,\uparrow} + c_{-k,\downarrow}^+ c_{-k,\downarrow} - 1)$$

$$- \sum_k (\Delta^* c_{-k,\downarrow} c_{k,\uparrow} + \Delta c_{k,\uparrow}^+ c_{-k,\downarrow}^+) + \Omega|\Delta|^2/U \tag{6.33}$$

with Δ and Δ^* parameters to be determined self-consistently.

The Bogolubov transformation used to diagonalize this quadratic form itself has to be linear and preserve momentum, spin and the anticommutation relations. It is only possible to satisfy all these conditions by violating particle-conservation:

$$c_{k,\sigma} \Rightarrow (\cos \vartheta_{k,\sigma})c_{k,\sigma} + (\sin \vartheta_{k,\sigma})c_{-k,-\sigma}^+ \,. \tag{6.34}$$

---

[m] A somewhat different (number conserving) analysis of the ground state solution to the nonlinear Hamiltonian in Eq. (6.31), D. C. Mattis and E. H. Lieb, *J. Math. Phys.* **2**, 602 (1961), leads to the same conclusions.

Still, the total number of electrons *is* conserved — but only on average. That is consistent with fixing μ (and not N) in a metal, and seemingly does not lead to any problems or paradoxes.[n]

Problem 6.5.   Prove that for the above transformation to be unitary and preserve the anticommutator algebra (6.30), $\vartheta_{k,\sigma}$ has to be real and odd *either* in k or in σ but not both: $\vartheta_{-k} = -\vartheta_k$ or $\vartheta_{-\sigma} = -\vartheta_\sigma$.

It is left as an exercise for the reader to derive the following, diagonal, form of (6.33) by optimizing $\vartheta_{k,\sigma}$.

$$H \Rightarrow \sum_{k,\sigma} \varepsilon_k \left(c_{k,\sigma}^+ c_{k,\sigma} - \frac{1}{2} \right) + \Omega |\Delta|^2/U \qquad (6.35)$$

where $\varepsilon_k = \sqrt{e_k^2 + |\Delta|^2} > 0$ is the quasi-particle's energy. Due to this pairing the total free energy achieves a minimum. It is,

$$F = -kT \sum_{k,\sigma} \log(1 + e^{-\beta \varepsilon_k}) - \sum_k \varepsilon_k + \Omega |\Delta|^2/U . \qquad (6.36)$$

The normal metal's quasiparticle spectrum has to be modified to take into account the "gap parameter" Δ. The effect, indicated by the dashed segment sketched in Fig. 6.2 on p. 150, is to eliminate the cusp and pin the gap squarely onto the Fermi surface.

One must optimize w.r. to the temperature-dependent gap Δ. Treating $\Delta \equiv \frac{U}{\Omega} \sum_k \langle c_{-k,\downarrow} c_{k,\uparrow} \rangle_{TA}$ and Δ^* as independent variables, one evaluates $\partial F/\partial \Delta^* = 0$ in (6.36) obtaining:

$$\Delta \equiv \frac{U}{\Omega} \sum_k \frac{\partial \varepsilon_k}{\partial \Delta^*} \{1 - 2f(\varepsilon_k)\} = \frac{U\Delta}{2\Omega} \sum_k \frac{\tanh \frac{1}{2}\beta \varepsilon_k}{\varepsilon_k} . \qquad (6.37)$$

Now, either $\Delta = 0$ or there exists a nontrivial, real, solution. In the thermodynamic limit, the sum becomes an integral. If $\Delta \neq 0$, divide (6.37) by Δ:

$$1 = U\rho(\mu) \int_0^{k\theta_D} de \frac{\tanh \frac{\beta}{2}\sqrt{e^2 + \Delta^2}}{\sqrt{e^2 + \Delta^2}} . \qquad (6.38)$$

At $T = 0$, $\tanh \frac{1}{2}\beta \varepsilon_k \to 1$ and this integral can be evaluated in closed form, yielding $\Delta(0) = k\theta_D \operatorname{cosech}(1/U\rho(\mu)) \approx 2k\theta_D \exp(-1/U\rho(\mu))$.

---

[n]In more sophisticated treatments a phase factor $e^{i\varphi}$ can be attached to each c^+ in (6.34).

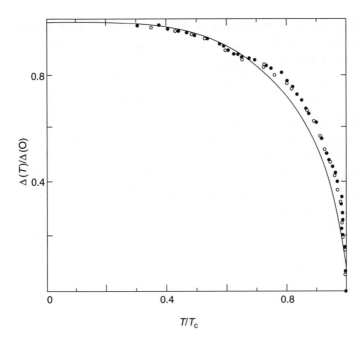

Fig. 6.6.   Dimensionless gap function in weak-coupling $(U\rho \to 0)$.
Example of BCS's "law of corresponding states": $\Delta(T)/\Delta(0)$ versus T/T_c.
Solid line: BCS theory, Eq. (6.36b). Points: experimental (for tin, Sn).

The exponential approximation is appropriate to the weak-coupling limit $U\rho \to 0$. For $T > 0$, $\Delta(T)$ is extracted numerically, as plotted in dimensionless form in Fig. 6.6.

The critical temperature T_c is extracted numerically from the same transcendental equation, assuming $\Delta_c \to 0$.

$$1 = U\rho(\mu) \int_0^{k\theta_D} de \frac{\tanh \frac{e}{2kT_c}}{e} = U\rho(\mu) \int_0^{\theta_D/2T_c} dx \frac{\tanh x}{x} . \qquad (6.39)$$

In the weak-coupling limit $(U\rho \to 0)$, $T_c \to 1.14\theta_D \exp(-1/U\rho(\mu))$. According to this, the dimensionless ratio $2\Delta(0)/kT_c = 3.528$. (In "strong-coupling" $(U \to \infty)$ this ratio is 4). The ratio $\Delta(T)/\Delta(0)$ is one of the universal functions of BCS's "law of corresponding states".

In addition to exhibiting zero electrical resistance, the superconductor is a cross between an ideal Fermi gas, i.e. an ordinary metal, and an insulator, given the gap that hinders the production of quasiparticles. Thus, features that depend on the *dos* at the *FS* are affected when a metal becomes superconducting. This phase transition can be first-order (as in decreasing the magnetic field from above to below $B_c(T)$ at $T < T_c$), or second-order (by lowering T to below T_c in zero field). Either way, numerous experiments confirmed the BCS predictions in some detail.

From among several well-known examples we mention *ultrasonic atten-uation*. There, the presence of a gap prevents damping by the Fermi sea through the production of a wake of low-energy quasiparticle pairs, therefore the propagation of low-frequency waves — again, a property of the atomic masses — is greatly enhanced. In *NMR* the energy gap affects the number of quasiparticles within $\pm\hbar\omega$ of the *FS* able to flip their spins and thus interact with the nuclear spins, decreasing them at low T/T_c but actually increasing their *dos* near T_c near "resonance", i.e. for $\hbar\omega \approx 2\Delta(T)$. To this list we should add "optical" absorption (really, microwave and far infrared absorption). As in an insulator, this becomes vanishingly small at frequencies $\hbar\omega < 2\Delta(T)$. But the salient feature of superconductors is the disappearance of all *dc* electrical resistance. How does the BCS theory deal with that? The following simplistic explanation, essentially a form of two-fluid theory, may suffice.[o]

Suppose a shift $\hbar\mathbf{q}$ in the origin of the Fermi sphere causes a current $\mathbf{J}$ to flow. We perform a Galilean transformation onto the moving coördinate system. With phonons effectively "frozen out" at low $T \ll \theta_D$, only elastic scattering mechanisms that break the translational invariance of the crystal, such as surfaces, grain boundaries and impurities, are capable of scattering a current once it flows. Consider the simplest example, that of a single point scatterer at the origin. Its Hamiltonian before and after the Bogolubov transformation is:

$$H' = \frac{g}{\Omega} \sum_{k,\sigma} \sum_{k'} c^{+}_{k,\sigma} c_{k',\sigma}$$

$$\Rightarrow \frac{g}{\Omega} \sum_{k,\sigma} \sum_{k'} [(\cos\vartheta_{k,\sigma}\cos\vartheta_{k',\sigma} + \sin\vartheta_{k,\sigma}\sin\vartheta_{k',\sigma}) c^{+}_{k,\sigma} c_{k',\sigma}$$

$$- (\cos\vartheta_{k,\sigma}\sin\vartheta_{k',\sigma}) c^{+}_{k,\sigma} c^{+}_{k',-\sigma} + H.C.] \tag{6.40}$$

Two distinct terms are seen on the *rhs*: elastic scattering (c^+c) terms and inherently inelastic pair creation/annihilation terms. Although the latter cannot damp a *dc* current without changing the energy, the elastic scattering

---

[o]More convincing explanations exist, but they can become mathematically rather in-volved; see Chapter 6 entitled "electrodynamics", in G. Rickayzen's thoughtful monograph, *Theory of Superconductivity*, Wiley, New York, 1965.

may (although at $T = 0$ *there are no quasiparticles present* and thus there are none to be scattered!) At finite T, thermal quasiparticles present can be scattered. Even though this scattering does not vanish identically, the super- and quasiparticle-currents flow in parallel. Thus, any superconducting current "shorts out" the normal currents below T_c, immediately upon the onset of a nonvanishing energy gap.

For it seems that, once created, the paired "bulk" current $\mathbf{J} = eN_{el}\hbar\mathbf{q}/m$ persists indefinitely at $T = 0$, there being no low-order mechanism whereby it could be elastically scattered by one-body perturbations such as (6.40). Indeed, in experimental studies the lifetimes of persistent currents, ultimately limited by higher-order processes, have been estimated to exceed thousands of years.

Exclusion of magnetic flux — the Meissner effect — is a related phenomenon that can be understood semiphenomenologically in the same manner.

In a magnetic field, each $\mathbf{p}_j \to \mathbf{p}_j - e\mathbf{A}/c$. Considering this to be as a local shift in the origin of the Fermi sea in the amount $\hbar\mathbf{q} = -e\mathbf{A}/c$, one uses the preceding arguments to obtain a current density in the amount $\mathbf{j} = -n_{el}e^2\mathbf{A}/mc$, where $n_{el} = N_{el}/\Omega$. The relevant Maxwell equation is $\nabla \times \mathbf{B} = -\frac{4\pi}{c}\mathbf{j}$. Combining the two one obtains a wave equation,

$$\nabla \times \nabla \times \mathbf{B} = -\nabla^2\mathbf{B} = -\frac{4\pi n_{el}e^2}{mc^2}.$$

Its solution, for the geometry where $\mathbf{B}$ is parallel to the metal's surface and decays with distance z into the bulk, is $\mathbf{B}(z) = \mathbf{B}(0)\exp-(z/\lambda_L)$, where $\lambda_L \equiv \sqrt{(mc^2/4\pi n_{el}e^2)}$ is the "London penetration distance".[p] Beyond this distance the magnetic field within the bulk of the superconductor is effectively zero. Thus, a metallic, superconducting ring of radius R will trap magnetic flux within its circumference; the amount of trapped flux Φ is quantized due to the requirement that $|q|$ in (6.31) = integer$/R = 2e|A|/\hbar c$.[q] The quantized trapped flux is therefore $\Phi \equiv B\pi R^2 = 2\pi|A|R = \hbar/e \times$ integer.

Once the surface magnetic field exceeds a critical magnitude $B_c(T)$, the energy of the excluded magnetic field is too high for the superconductor to sustain and the normal state is restored. This is a first-order phase transition, previously analyzed in Sec. 3.11 using generic Clausius–Clapeyron type arguments.

---

[p]F. London and H. London, *Proc. Royal Soc.* (*London*) **A149**, 71 (1935) and *Physica* **2**, 341 (1935). For details, see Rickayzen, *op. cit.* (footnote o).
[q]We use $2e$ in this formula, as it is the charge of a bound Cooper pair.

6.14. Contemporary Developments in Superconductivity

The interest surrounding the BCS theory stimulated numerous discoveries, including the pair-breaking effects of magnetic impurities that can significantly decrease the energy gap Δ and to a lesser degree, T_c. The discovery of type II superconductors that magnetic fields could penetrate by creating arrays of *vortices* with non-superconducting cores added to the general excitement. Inhomogeneities of this type became understood with the aid of semiphenomenological Landau-Ginsberg equations, once they had been mated to the BCS theory.[o]

The Josephson effect and numerous other tunneling phenomena soon became major concerns,[r] although the real push was on to raise T_c to where practical applications might ensue. For a period of 3 decades all known superconducting materials were alloyed and mixed with one another in a futile attempt to raise T_c above some $25°K$. In fact, a number of theoretical speculations (too erudite to repeat) purported to show that $T_c = 30°K$ could never be exceeded by the mechanism of the electron-phonon interaction alone.

The situation changed dramatically in late 1986 and the Nobel prize was awarded to J. Bednorz and K. Müller shortly thereafter for their discovery of the first "high-temperature superconductor" — doped lanthanum copper oxide, a layered mineral belonging to the perovskite family, exhibiting type II superconductivity up to $T = 35$ K. Numerous other minerals, all containing layers of the two-dimensional spin-1/2 antiferromagnet CuO_2, were soon developed in a frantic search for successively higher T_c. Interestingly, the normal conductivity in these materials is highly *anisotropic*: highly conductive in the *ab* plane, they are approximately semiconductors along the perpendicular crystallographic *c*-axis. On the other hand the superconducting phase is quasi-*isotropic*, possibly owing to Josephson tunneling between planes that is lacking in the normal phase.

Critical temperatures upwards of 160 K have already been attained (or at least, reported) and at the date of writing, some hope remains that a room-temperature superconductor will be found. It is not far-fetched to say that such a discovery would soon revolutionize contemporary electronics, electrical engineering and perhaps all of technology.

For a variety of reasons: the absence of significant isotope effect, carrier mobility too high to be compatible with strong electron-phonon coupling,

---

[r]Brian Josephson won the 1973 Nobel prize for his 1962 predictions of oscillations and tunneling associated with pairs and quasiparticle tunneling; see D. Langenberg, D. Scalapino and B. Taylor, *Scientific American* **214**, 30 (1966).

etc., it has been suspected that the pairing interactions in the copper oxides are mediated by antiferromagnetic exchange interaction among the spins of the charge carriers. Just such a mechanism is incorporated in the "t-J model" of the electron gas, itself loosely based on a mechanism called "superexchange" that is known to be operative in insulating magnetic metal oxides including CuO_2. A related spin-wave exchange mechanism is thought to cause pairing and superfluidity in the fermion liquid ^3He at extremely low temperatures, $O(10^{-3}$ K$)$. But at the date of writing, it is still not understood why high-temperature superconductivity seemingly occurs *only* in materials containing *two-dimensional CuO_2* and not in other antiferromagnets or geometries.

Indeed, at the time of writing, the mechanism whereby charge carriers in CuO_2 attract and form Cooper pairs remains controversial and mysterious. Also controversial is the nature of the energy gap in high-temperature superconductors; the concensus has it that it exhibits a d-wave symmetry — vanishing along two crystal directions — instead of being constant along surfaces of constant energy, as it is in the "classic" low-T_c superconductors. Possibly the lack of definitive answers is due not so much to the lack of experimental data, but (to quote Feynman once again) to the fact that we *still* "...haven't got enough imagination".[h]

Chapter 7

Kinetic Theory

7.1. Scope of This Chapter

Presently we shall describe attempts to extend thermodynamical concepts beyond thermodynamic equilibrium. Among applications we count the spontaneous approach to equilibrium starting from arbitrary initial conditions, a formulation of transport theory and a theory of the propagation and attenuation of collective modes (e.g. sound waves) in gases. In a subsequent chapter we expand our technical arsenal with the aid of boson and spin operators, Green functions and elementary quantum field theory.

Kinetic theory is an approximation justified only if interactions among "ideal gas" constituents are weak or their collisions infrequent, sufficiently so that neither significantly affects the thermodynamic functions. But ideal gases have $3N$ constants of the motion and are incapable of *any* approach to thermodynamic equilibrium! Therefore one anticipates that weak interactions induce nontrivial changes in a system. It is necessary to determine *how weak* interactions have to be before we can unequivocally use kinetic theory.

In the systems under consideration the constituent "particles" can be any of various atomic or molecular species, bosons, fermions, or even normal modes, etc. The rearrangements can result from *scattering*, typically $\mathbf{p}_1$, $\mathbf{p}_2 \Leftrightarrow \mathbf{p}_1 + \mathbf{q}$, $\mathbf{p}_2 - \mathbf{q}$, or from a chemical reaction (actually, a generalized form of scattering), such as $2H_2 + O_2 \Leftrightarrow 2H_2O$.

Defining, deriving and then using the notion of "detailed balance" (the absence of any net flow or transitions in thermodynamic equilibrium) we shall show that the very *form* of the collision integral determines the "statistics" of colliding particles and can reveal whether they are bosons, fermions, or classical entities! This is the approach pioneered by Einstein in his derivation of Planck's distribution law for photons in blackbody radiation. We start by determining the *direction* of flow (heat flow, particle flow, electrical current flow, etc), by recapitulating arguments derived from the Second Law alone.

Then with the aid of the *collision integral* we show the approach to equilibrium to be monotonic, first by proving then by using Boltzmann's "H-theorem", and deriving a "Master Equation". Finally, combining "streaming" terms with the collision integral we establish *Boltzmann equation* and solve it in selected examples, deriving *Ohm's law* of electrical conduction on the one hand and sound propagation and attenuation in gases on the other.

7.2. Quasi-Equilibrium Flows and the Second Law

Connect two containers *via* a slow-moving piston while isolating them from the outside, such that $V_1 + V_2 = V$ remains constant and $E = E_1 + E_2$ remains constant as well. According to the Second Law the entropy in thermodynamic equilibrium always tends to a maximum, therefore if there is a change $d\mathscr{S}$ it must be > 0. Now,

$$d\mathscr{S}_1 = \frac{dE_1}{T_1} + \frac{p_1 dV_1}{T_1}, \quad d\mathscr{S}_2 = \frac{dE_2}{T_2} + \frac{p_2 dV_2}{T_2}. \tag{7.1}$$

Therefore $d\mathscr{S}_1 + d\mathscr{S}_2 > 0$ implies,

$$\left\{ \frac{T_2 - T_1}{T_2 T_1} \right\} dE_1 + \left\{ \frac{p_1 T_2 - p_2 T_1}{T_2 T_1} \right\} dV_1 > 0. \tag{7.2}$$

As dE_1 and dV_1 are arbitrary, each term must independently be > 0. Thus, if $p_1/T_1 = p_2/T_2$, the temperature difference $T_2 - T_1$ must have the same sign as dE_1, i.e. if container #1 is gaining energy container #2 must be at a higher temperature. If the temperatures are equal but the pressures are not, $p_1 - p_2$ has the sign of dV_1, i.e. the piston moves from the region of higher to that of lower pressure. A similar argument, using $N_1 + N_2 = $ constant, governs the flow of particles from higher to lower μ. Finally, the flow of dissimilar particles entails the additional "entropy of mixing".

The following example, previously treated in Problem 4.1, should serve: let N particles of isotope #1 be in container of volume V at p, T while N particles of isotope #2 are under identical conditions. Abruptly, the barrier separating the two volumes is removed. It is desired to show that asymptotically, after conditions of thermodynamic equilibrium have been restored, each species will occupy both containers fully. *The proof*: Eq. (4.4) specifies the free energy F of an ideal gas, $\mathscr{S} = -\partial F/\partial T$ its entropy. Because N and T are unchanged if both constituents occupy the joint volume $2V$,

$\Delta F_{\text{total}} = (F_{1,\text{final}} + F_{2,\text{final}}) - (F_{1,\text{initial}} + F_{2,\text{initial}}) = -2kTN \log 2$. Hence $\Delta \mathscr{S} = 2kN \log 2 > 0$ is the entropy of mixing.

Exercise for the reader: in the event the two species had been strictly indistinguishable, show that the use of the correct Gibbs factors would have resulted in $\Delta F_{\text{total}} = 0$ and the same mixing would have been thermodynamically irrelevant or inconsequential.

7.3.   The Collision Integral

Although it might seem obvious that thermodynamic equilibrium is attained only when the entropy (read, "probability"), is maximized, the process of redistribution of the individual particles' energies *via* molecular collisions is rather complicated. In fact there is no general, rigorous, proof that the sequence of these collisions is teleological, i.e. so organized as to result in an optimal state. Still, no sensible person disputes the usefulness of the *ergodic hypothesis* — an attractive and commonsensical conjecture that we discuss later. In any event, for any such study a detailed understanding of molecular collisions is essential.

If their potentials are short-ranged two particles are able to exchange some momentum (i.e. "collide") *only* if b, the distance of nearest-approach (*aka* impact parameter), shown in the figure, does not exceed the range of the potential. A good, intuitive, example is that of two billiard balls of radius a that do not collide unless their trajectories bring their centers closer than $b = 2a$.

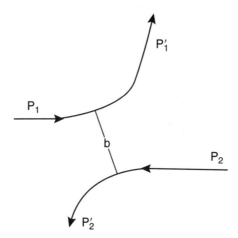

Fig. 7.1.   A scattering event, indicating the impact parameter b.

Thus there is a correlation between the particles' positions and their collisions. A number of conservation laws govern the scattering events, among which conservation of momentum and energy play vital rôles.

Quantum theory provides a framework in which to compute transition probabilities *per* unit time. In low-order perturbation theory it yields what is known as the "Born approximation". From it one finds the incremental *probability* of a collision to be proportional to 3 independent factors:

- a *rate* K = collision cross-section × relative velocity,
- a "statistical" factor Ψ related to the generalized Pauli Principle,
- and the (infinitesimal) passage of time dt.

Delta functions enforce the conservation laws. In the case of collisions among identical particles the incremental probability is:

$$dP = K(1,2|1',2')\Psi(1,2|1',2')\delta(\mathbf{p}_1+\mathbf{p}_2-\mathbf{p}_1'-\mathbf{p}_2')\delta(\varepsilon_1+\varepsilon_2-\varepsilon_1'-\varepsilon_2')dt\,. \quad (7.3)$$

$K > 0$ is a *kinematical* factor that is generally the same for fermions as for bosons. It is symmetric under the interchange of the incoming (colliding) particles[a] 1,2 and under the interchange of the outgoing particles $1',2'$, and symmetric under the interchange of the incoming *pair* (1,2) wih the outgoing pair $(1',2')$.

The factor $\Psi(1,2|1',2') = -\Psi(1',2'|1,2)$ is *antisymmetric* under the interchange of the incoming pair 1,2 with the outgoing pair $1',2'$. It expresses the principle of *detailed balance*. Let us write it more explicitly as: $\Psi(1,2|1',2') = \Psi_+(1,2|1',2') - \Psi_+(1',2'|1,2)$. The fundamental *nature* of the particles — that which determines whether they are *bosons, fermions* or whatever — is *entirely contained* within Ψ_+.

In the generalized transition matrix element ("*ME*") that is used to compute Ψ_+ there is a destruction operator c for each of the incoming particles and a creation operator c^+ for each of the outgoing particles. In the special case of initially free particles (appropriate for the study of the quasi-ideal gas), the operators are labeled to take conservation of total momentum explicitly into account. Take as an example the collision shown in Fig. 7.1, for which the *probability amplitude* for a transition from an initial state $|i\rangle$ to a final state $\langle f|$ is

$$ME = \langle f|c^+_{p_1+q}c^+_{p_2-q}c_{p_2}c_{p_1}|i\rangle\,,$$

[a]If each particle carries spin and/or other identifying labels, "1" stands for the complete set of labels of the first particle, "2" for the second, etc.

The probability is the *square* of the *modulus* (absolute value) of the probability amplitude, $|ME|^2$. To find the *form* of the answer, sum $|ME|^2$ over all states $|f\rangle$ to obtain[b] a quantity we'll define as M^2:

$$M^2 = \langle i | c_{p_1+q} c_{p_1+q}^+ c_{p_2-q} c_{p_2-q}^+ c_{p_2}^+ c_{p_2} c_{p_1}^+ c_{p_1} | i \rangle \, .$$

The thermodynamic average of M^2 performed in the neighborhood where the collision takes place, at time t, yields Ψ_+. That is,

$$\Psi_+ \equiv \langle M^2 \rangle_{TA} = \langle (1 \pm n_{p_1+q})(1 \pm n_{p_2-q}) n_{p_2} n_{p_1} \rangle_{TA} \, . \qquad (7.4)$$

The upper sign (+) is for bosons, the lower sign for fermions. This result is perfectly intuitive although it deserves some discussion.

In the *classical* (dilute) limit, $\langle n_p \rangle_{TA} = f(p) \ll 1$; in thermal equilibrium f must reduce to precisely the Maxwell–Boltzmann distribution discussed in Eqs. (4.12)–(4.14). To leading order in the density we may then set $\Psi_+ = f(p_1) \cdot f(p_2)$, proportional to the density of particles in each of the incoming channels.[c] Regardless of the ultimate value of f, in setting[d] $\langle n_1 n_2 \rangle$ specifically $= f_1 \cdot f_2$ (rather than some correlation function $g(p_1, p_2)$), we have implicitly assumed the average occupancy of p_1 to be statistically independent of the average occupancy of p_2 prior to the collision in question, i.e. we have approximated the expectation value by that in an ideal gas. The true effects of this collision (one of which is to correlate the states), have to be be found *post-hoc* by different means.

For *fermions*, a similar decoupling leads to,

$$\Psi_+ = f(p_1) \cdot f(p_2) \times (1 - f(p_1 + q)) \cdot (1 - f(p_2 - q)) \, ,$$

with the two additional factors clearly reflective of Pauli's exclusion principle in some thermodynamically averaged way.

For *bosons*:

$$\Psi_+ = f(p_1) \cdot f(p_2) \times (1 + f(p_1 + q)) \cdot (1 + f(p_2 - q)) \, .$$

---

[b]Using the "completeness" theorem of quantum mechanics.

[c]This formula was surmised long before the advent of quantum mechanics. Note that here f is as yet unspecified and may be a function of time and position.

[d]This decoupling, frequently called "hypothesis of molecular chaos" and originally denoted "Stosszahlansatz" by Boltzmann, presupposes that previous collisions had not set up correlations among the various channels, i.e. that collisions are just small and uncorrelated, like RW steps in some phase space.

In the factors $(1 + f)$, the constant "1" represents "spontaneous" emission and "f" the "stimulated" emission.[e]

Because dP is to be the *net* change in probability during the time interval dt, the reverse process has to be subtracted. According to

The principle of detailed balance,

In thermodynamic equilibrium the rate at which $\mathbf{p}_1, \mathbf{p}_2$ convert into $\mathbf{p}_1 + \mathbf{q}$, $\mathbf{p}_2 - \mathbf{q}$ must equal *precisely* the rate at which $\mathbf{p}_1 + \mathbf{q}$, $\mathbf{p}_2 - \mathbf{q}$ convert back into $\mathbf{p}_1$, $\mathbf{p}_2$; this holds for *each* pair $\mathbf{p}_1, \mathbf{p}_2$ and for *each* value of $\mathbf{q}$.

This notion is based solely on the reversibility of the microscopic equations of motion.[f] The very *form* (antisymmetric) chosen for Ψ guarantees that, regardless of the details of the scattering process, the population in each and every channel remains absolutely stationary in thermodynamic equilibrium. This form is believed to be exact in the limit that the scattering is either very weak or infrequent.

We next examine the approach to equilibrium of a classical, dilute, gas of particles, derive the "classical" Maxwell–Boltzmann distribution function and prove Boltzmann's notorious "H-theorem".

7.4.   Approach to Equilibrium of a "Classical" Non-Ideal Gas

From the preceding we can compute the rate at which the occupancy of momentum state p decreases due to collisions, as

$$\left. \frac{\partial f(p_1, t)}{\partial t} \right|_{\text{coll}} = - \int d^3 p_2 \int d^3 p_1' \int d^3 p_2'$$

$$\times K(p_1, p_2 | p_1', p_2') \delta(p_1 + p_2 - p_1' - p_2') \delta(\varepsilon_1 + \varepsilon_2 - \varepsilon_1' - \varepsilon_2')$$

$$\times \{ f(p_1, t) f(p_2, t) - f(p_1', t) f(p_2', t) \} . \tag{7.5}$$

In writing this expression we tacitly assumed $f(p, t)$ to be spatially homogeneous and K to have become averaged over all $\mathbf{r}_1, \mathbf{r}_2, \mathbf{r}_{1'}, \mathbf{r}_{2'}$. This simplification is not always justified and should be reëxamined for processes occuring at surfaces or interfaces.

Integrating over p_1 instead of p_2 would yield $\partial f(p_2, t)/\partial t$ but no new information, as p_1 and p_2 range over an identical set of values. Owing to its

---

[e]Paraphrasing Einstein's derivation of Planck's Law (governing the distribution of photons in blackbody radiation), to which we shall return in due course.
[f]Or, more technically, on the Hermitean nature of the scattering Hamiltonian.

symmetry, Eq. (7.5) conserves particles, kinetic energy and all three components of total momentum. For example,

$$0 = \int d^3 p_1 \left. \frac{\partial f(p_1, t)}{\partial t} \right|_{\text{coll}} = \left. \frac{\partial N}{\partial t} \right|_{\text{coll}} \tag{7.6}$$

in which the first equality follows from the antisymmetry of Ψ, the curly bracket and the second, from interchanging $\partial/\partial t$ with the integration. The 4 other conservations laws can be proved in a similar manner.

Problem 7.1. Use Eq. (7.5) to prove conservation of total *energy* $\partial E/\partial t|_{\text{coll}} = 0$ and conservation of each of the 3 components of total *momentum* $\partial \mathbf{P}/\partial t|_{\text{coll}} = 0$, assuming the ideal-gas definitions $E(t) = \int d^3 p \varepsilon(p) f(p, t)$ and $\mathbf{P}(t) = \int d^3 p \mathbf{p} f(p, t)$ to hold.

The entropy of this quasi-ideal gas is, of course, the key to the approach to equilibrium. As originally given in Eq. (1.21), the entropy in each channel p is related to $\langle \log n_p! \rangle_{TA}$. Boltzmann selected the closely related quantity $f(p) \log f(p) - f(p)$ with which to construct his "H-function",

$$H(t) \equiv \int d^3 p [f(p, t) \log f(p, t) - f(p, t)]. \tag{7.7}$$

He then used Eq. (7.5) with which to evaluate the time derivative:

$$\left. \frac{dH(t)}{dt} \right|_{\text{coll}} = -\prod_{i=1}^{4} \int d^3 p_i K(p_1, p_2 | p_1', p_2') \delta(p_1 + p_2 - p_{1'} - p_{2'})$$

$$\times \delta(\varepsilon_1 + \varepsilon_2 - \varepsilon_{1'} - \varepsilon_{2'}) \{ f(p_1, t) f(p_2, t) - f(p_1', t) f(p_2', t) \} \log f(p_1, t)$$

writing p_1 for p.

Note that with the exception of the log term, all factors in the 12-dimensional integration are antisymmetric under the interchange of the incoming with the outgoing pairs. We can symmetrize the integrand without affecting the outcome of the integration:

$$\left. \frac{dH(t)}{dt} \right|_{\text{coll}} = -\frac{1}{4} \prod_{i=1}^{4} \int d^3 p_i K(p_1, p_2 | p_1', p_2') \delta(p_1 + p_2 - p_{1'} - p_{2'})$$

$$\times \delta(\varepsilon_1 + \varepsilon_2 - \varepsilon_{1'} - \varepsilon_{2'}) \{ f(p_1, t) f(p_2, t) - f(p_1', t) f(p_2', t) \}$$

$$\times \log \frac{f(p_1, t) f(p_2, t)}{f(p_1', t) f(p_2', t)}. \tag{7.8}$$

The { } terms × the log terms now have the form $\{x - y\} \log(x/y)$. The reader will verify that this quantity is ≥ 0 for any x, y real numbers. The delta functions and K are inherently ≥ 0, hence the integrand is everywhere non-negative and the integral ≥ 0.

If it is a thermodynamic function, H cannot decrease forever and $\partial H/\partial t$ must ultimately vanish when thermal equilibrium is established. Eq. (7.8) shows this can occur *iff* the integrand vanishes identically at all values of the arguments, i.e. *iff* asymptotically,

$$f(p_1, \infty)f(p_2, \infty) = f(p_1', \infty)f(p_2', \infty)$$

i.e.

$$\log f(p_1, \infty) + \log f(p_2, \infty) = \log f(p_1', \infty) + \log f(p_2', \infty).$$

Now, scattering occurs "on the energy shell" and all three components of total momentum are conserved. It follows that the conditions above, which are the conditions for thermodynamic equilibrium, *are* satisfied by $\log f(p, \infty) = \mathbf{a} \cdot \mathbf{p} + b\varepsilon(p) + c$, where $\mathbf{a}, b$, and c constitute a set of 5 constants coresponding to 5 conservation laws: 3 for momentum, 1 for energy and 1 for constants, e.g. μ.

In a system at rest there is no preferred direction, hence we must choose $\mathbf{a} = 0, b = -\beta$ and, finally, c as the constant required to normalize f to N. These choices ensure that asymptotically, f is just the equilibrium Maxwell–Boltzmann distribution function. The actual values of β (a function of b) and N (a function of b and c) in the system depend on initial conditions and cannot be derived from the calculation. One may also ask whether, after including explicit time diferentiations and supplementing the collision term in Eq. (7.8) by adding $\partial H/\partial t$ to it, we could choose the asymptotic $\mathbf{a}, b$ and c as functions of t? The answer is found in the following Problem.

Problem 7.2.   Using the laws of conservation of energy and momentum to somewhat "sharpen" the conclusions of Problem 7.1, prove that the 5 parameters in the *asymptotic* distribution function found above must be constants in time — that $\mathbf{a}, b, c$, *cannot* be periodic or aperiodic functions of t.

7.5.   A New Look at "Quantum Statistics"

For *fermions* the curly bracket in Eqs. (7.5) and (7.8) is replaced by
$$\{\Psi_+(1, 2|1', 2') - \Psi_+(1', 2'|1, 2)\} = \{f(p_1, t) \cdot f(p_2, t) \cdot (1 - f(p_1 + q, t)) \cdot (1 - $$

$f(p_2-q,t))-f(p_1+q,t)\cdot f(p_2-q,t)\cdot(1-f(p_1,t))\cdot(1-f(p_2,t))\}$. This bracket
vanishes asymptotically only when $\log \Psi_+(1,2|1',2') = \log \Psi_+(1',2'|1,2)$. We
rewrite the present result using the f's, and collect terms:

$$\log \frac{f(p_1,t)}{1-f(p_1,t)} + \log \frac{f(p_2,t)}{1-f(p_2,t)}$$

$$= \log \frac{f(p_1',t)}{1-f(p_1',t)} + \log \frac{f(p_2',t)}{1-f(p_2',t)}. \tag{7.9}$$

The asymptotic solution occurs, presumably, what $t \rightarrow \infty$. It is
$\log \frac{f(p,\infty)}{1-f(p,\infty)} = \mathbf{a}\cdot\mathbf{p}+b\varepsilon(p)+c$, with the constants having the same meanings
as before. We identify $b = -\beta$, $c = \beta\mu$, and set $\mathbf{a} = 0$. The result for a system
at rest is,

$$f(p,\infty) \equiv f(p) = \frac{1}{e^{\beta(\varepsilon(p)-\mu)}+1} \qquad (fermions) \tag{7.10a}$$

For *bosons* the reader will verify that,

$$f(p,\infty) = f(p) = \frac{1}{e^{\beta(\varepsilon(p)-\mu)}-1}. \qquad (bosons) \tag{7.10b}$$

Quite generally the vanishing of the $\{\Psi\}$ bracket signifies both *thermal
equilibrium* and *detailed balance*. Amazingly, we have used collisions to determine the distribution functions of the ideal gases! This is only possible
because of our neglect of correlations.

Let us review how nearly 100 years ago Einstein deduced Planck's law
from detailed balance. Suppose N_{exc} is the number of atoms capable of emitting radiation $\hbar\omega$ and N_0 the number that remain after the emission (with
$N_{\text{total}} = N_{\text{exc}}+N_0$). The relevant $\Psi = \{N_{\text{exc}}(1+f)-N_0f\}$ vanishes when the
number of photons is $f = [N_0/N_{\text{exc}} - 1]^{-1}$. If the emitting atoms are in thermal equilibrium, $N_0/N_{\text{exc}} = \exp \hbar\omega$. From this follows *Planck's law*, precisely
as derived from "first principles", i.e. statistical mechanics, in Eq. (5.17).
The vanishing of the $\{\Psi\}$ bracket signifies that the radiation is, on average,
in thermal equilibrium with the excited and ground-state atoms; although
photons are constantly emitted they are being absorbed at precisely the rate
required to keep their number and the fraction of excited atoms constant.
Fluctuations ("noise") are ignored in this formulation, as are correlations.

A laser can emit at the given frequency only to the extent that the number
of excited atoms exceeds the thermal value, $N_0 \exp -\hbar\omega$, and therefore some
dynamic, non-equilibrium mechanism has to be found to "feed" this excited
state. Equations that govern the ebb and flow of *probability distributions* fall

under the general category of "Master Equations", of which the simplest example, radioactive decay, is treated next.

7.6. Master Equation: Application to Radioactive Decay

Master Equation is the name given to any equation regulating the flow of probabilities in a stochastic process. A master equation can be used to formulate transport theory or fluctuations. It can allow the detailed study of a diffusive chemical reaction. A master equation is at the basis of any satisfactory theory of turbulence that allows the study of fluctuations as well as the evolution of averages of quantities.

In this simplest of applications, we examine the decay of N nuclei in the absence of any correlations among them. Even so, the law of large numbers allows the solution of the master equation to predict properties of the ensemble that the study of a single radioactive nucleus would not be able to reveal. We shall wish to know $P(n|t)$, the probability that n out of the initial N nuclei have decayed after time t has elapsed.

Let me emphasize what we are *not* doing: we do not look just for what is the average number of decayed nuclei after time t, a question that is immediately resolved by solving the following trivial equation, $d\langle n \rangle / dt = (N - \langle n \rangle)/\tau$, with τ the empirical $1/e$ lifetime of an individual nucleus. That would be the "Boltzmann equation"type approach. Subject to the initial condition $n(0) = 0$, is (by inspection) $\langle n \rangle (t) = N(1 - \exp{-t/\tau})$, a monotonically increasing function whose asymptotic value is N. But as in some examples of the RW of Chapter 1, such information is incomplete. It does not help us determine the time dependence in $\langle n^2 \rangle (t)$ nor the time-dependence of the related quantity $\sigma^2(t) = \langle [n - \langle n \rangle (t)]^2 \rangle$ that measures the "noise" in the distribution as a function of t. The key point is the presence everywhere of $\langle \cdots \rangle$, *requiring us to find a probability* over which to average.

The equation satisfied by $P(n|t)$ is what we seek, the "master equation" in the present context. It is, quite intuitively,

$$dP(n|t) = \left\{ (N - n + 1)P(n-1|t) - (N-n)P(n|t) \right\} \frac{dt}{\tau} \qquad (7.11)$$

and it expresses the change in heights of the histograms P for $n = 0, 1, \ldots, N$ in the brief time interval dt.

In words: there is a probability $P(n|t)$ that precisely n have decayed. It *decreases* by dP_1 if an additional decay occurs in time dt; the probability of this happening is $\propto (N - n)P(n|t)/\tau$. Conversely, if only $n - 1$ had decayed at time t, the decay of a single additional nucleus in time $dt(\propto (N - n + 1)$ $P(n-1|t)/\tau)$ *boosts* $P(n|t)$ by dP_2. The incremental time dt must be chosen $\ll \tau$ in order that the probability of any *two* nuclei decaying within dt tends to zero and does not affect the conclusion. Then $dP = dP_2 - dP_1$.

A complete solution of this difference-differential equation requires obtaining $P(n|t)$ for all integers $n = 0, \ldots, N$ at all times $t > 0$, subject to the initial condition: $P(n|0) = \delta_{n,0}$ and normalization $\Sigma_n P(n|t) = 1$. It is not easy to solve this equation straightforwardly and we shall require a "trick". But it is trivial to show that if P is a solution to Eq. (7.11) and normalized at $t = 0$, it *remains* normalized subsequently. [*Exercise for the reader*: using Eq. (7.11) to evaluate $d/dt\,\Sigma_n P(n|t)$, show that two sums cancel and $\Sigma_n P(n|t)$ is therefore a constant, equal to its initial value at $t = 0$.]

Difference equations are typically solved using Fourier transformation. Here, instead of exp *in* φ, we introduce a new variable z with which to construct the following *generating function*:

$$G(z,t) \equiv \sum_{n=0}^{N} z^{N-n} P(n|t), \quad \text{therefore, } P(N - n|t) = \frac{1}{n!}\frac{\partial^n}{\partial z^n}G(z,t)\bigg|_{z=0}.$$
(7.12)

It is subject to the initial condition $G(z,0) = z^N$ [reader: why?] and, according to (7.11), satisfies the *p.d.e.*:

$$\frac{\partial G(z,t)}{\partial t} = \frac{1}{\tau}\sum_{n=0}^{N} z^{N-n}\{(N - n + 1)P(n - 1|t) - (N - n)P(n|t)\}$$

$$= \frac{-1}{\tau}(z - 1)\frac{\partial G(z,t)}{\partial z}.$$
(7.13)

This equation is satisfied by *any* real function $G(z,t) = G(\log(z-1)-t/\tau)$. Initially, $G(z,0) = z^N = [e^{\log(z-1)} + 1]^N$. Therefore at finite t one may infer:

$$G(z,t) = [e^{\log(z-1)-t/\tau} + 1]^N = e^{-Nt/\tau}[z - 1 + e^{t/\tau}]^N$$

$$= e^{-Nt/\tau} \sum_{n=0}^{N} \binom{N}{n} z^{N-n}(e^{t/\tau} - 1)^n.$$
(7.14)

Equating powers of z here and in (7.12) yields the normalized P's,

$$P(n|t) = e^{-Nt/\tau}\binom{N}{n}(e^{t/\tau} - 1)^n.$$
(7.15)

For the purpose of calculating $\langle n \rangle$, $\langle n^2 \rangle$, ..., one does not even need to know the P's and can just use G directly as follows:

$$\langle (N-n)^p \rangle = \left(z \frac{\partial}{\partial z} \right)^p G(z,t) \bigg|_{z=1} , \qquad (7.16)$$

where G is expressed explicitly in (7.14).

Problem 7.3. (A) Compute and plot $\sigma^2(t) = (\langle n^2 \rangle - \langle n \rangle^2)/N$.

(B) Compute the maximum $P(n|t)$ at fixed t; call this $n_{\max}(t)$. Sketch or plot $n_{\max}(t)$, compare with $\langle n \rangle(t)$.

(C) Supposing a lethal dose of radiation were determined to be $n_{\text{leth}} = Ne^{-4}$ for $N = 10^4$. Plot the integrated probability of being exposed to a lethal dose (or greater!) as a function of t/τ.

Problem 7.4. The initial condition under which this model was solved was given as $P(n,0) = \delta_{n,0}$. But typically we know only the total number N of nuclei at $t = 0$, as some of them may have decayed previously. Absent any other information except that their average age $>>> \tau$ at $t = 0$, show that the statistically *most probable initial condition* $P(n,0)$ is some sort of exponential or Poisson distribution, and find what it is.

7.7. Boltzmann Equation

As we saw, the motion of particles is influenced by their collisions with other particles although their *net* migration is mainly influenced by applied forces and by density gradients. Upon combining the latter, the so-called "streaming terms", with the collision integral, we obtain the *Boltzmann equation*.

The Boltzmann equation ("BE") regulates the *average* flow of particles, hence the density, pressure, and other thermodynamically averaged properties of the dilute gas, but not fluctuations. Because it is so economically and persuasively formulated this equation is frequently extended beyond its strict domain of legitimacy.

Aside from the transport of conserved fermions the BE is also used to study the transport of heat by phonons (a typical example of non-conserved bosons), and the transport properties of mixed fermion and boson systems. The same equation is used to study *collective modes*, such as sound waves in an ordinary neutral, monatomic, dilute gas, and to compute *transport parameters* (electrical conductivity, Hall coefficient, thermal conductivity, etc.), in systems where scattering plays only a secondary rôle.

We first set up the BE, treat transport in the following sections and turn to the collective modes after that. Recall the *Liouville Equation* of classical dynamics, governing the motion of points (either their density $n(\mathbf{r}, \mathbf{p}, t)$ or their probability) through phase space. With $\langle n(\mathbf{r}, \mathbf{p}, t) \rangle \equiv f(\mathbf{r}, \mathbf{p}, t)$, the effects of spatial *drift*, of external *forces*, and of any explicit time-dependence on f are all collected in the Liouville operator $\mathbf{L} = \{ \partial/\partial t + \mathbf{v} \cdot \nabla_r + \mathbf{F} \cdot \nabla_p \}$:

$$\partial f(\mathbf{r}, \mathbf{p}, t)/\partial t|_{\text{stream}} = \{ \partial/\partial t + \mathbf{v} \cdot \nabla_r + \mathbf{F} \cdot \nabla_p \} f(\mathbf{r}, \mathbf{p}, t) \qquad (7.17\text{a})$$

where $\mathbf{v} = \mathbf{p}/m = d\mathbf{r}/dt$ for particles of mass m and $\mathbf{F} = d\mathbf{p}/dt$. The average number of particles in an infinitesimal 6-dimensional cube $d^3r \, d^3p$ is $f(\mathbf{r}, \mathbf{p}, t) d^3r \, d^3p$ and Eq. (7.17a) gives the rate at which it increases. However, the streaming of particles into $d^3r \, d^3p$ is countered by collisions. According to Eq. (7.5), and assuming spatial homogeneity, this rate is:

$$\partial f(\mathbf{r}, \mathbf{p}, t)/\partial t|_{\text{coll.}} = - \iiint d^3 p_2 d^3 p_1' d^3 p_2' K(p, p_2 | p_1', p_2')$$

$$\times \, \delta(\mathbf{p} + \mathbf{p}_2 - \mathbf{p}_{1'} - \mathbf{p}_{2'}) \delta(\varepsilon + \varepsilon_2 - \varepsilon_{1'} - \varepsilon_{2'})$$

$$\times \, \{ f(p, t) f(p_2, t) - f(p_1', t) f(p_2', t) \}$$

$$\equiv -C_2 \{ f(\mathbf{r}, \mathbf{p}, t) \} \qquad (7.17\text{b})$$

defining the two-body collision integral as a functional operator C_2 acting on f. It conserves energy and momentum. Other, distinct, collision integrals need to be introduced for inelastic processes and for scattering processes at impurities and at surfaces in which momentum is not conserved.

The *Boltzmann equation* ("BE") combines (7.17a) and (7.17b). With $df/dt|_{\text{net}} = \mathbf{L}\{ f(\mathbf{r}, \mathbf{p}, t) \} + C_2 \{ f(\mathbf{r}, \mathbf{p}, t) \} = 0$ for particle conservation,

$$\partial f(\mathbf{r}, \mathbf{p}, t)/\partial t|_{\text{stream}} = \partial f(\mathbf{r}, \mathbf{p}, t)/\partial t|_{\text{coll}}. \quad \text{(BE)} \qquad (7.18)$$

After some time has elapsed this equation may acquire an asymptotic solution $f(\mathbf{r}, \mathbf{p})$ that is independent of time, possibly sustaining a constant flow of current. It is tempting to conjecture that the collision term drives the distribution toward a steady-state of optimal entropy production just as it does in equilibrium (cf. Boltzmann's H-Theorem). A time-independent current-carrying solution cannot correspond to thermodynamic equilibrium because — essentially by definition — in equilibrium, no currents can flow. But a constant solution to (7.18) is compatible with a new and different state of matter denoted the *steady-state*.

One stretches the notion of "steady-state" to encompass cases where the applied forces $\mathbf{F} \propto u(t)$, periodic functions of time. For example, if

$f(\mathbf{r}, \mathbf{p}, t)$ becomes asymptotically synchronous with the applied force, i.e. if $f(\mathbf{r}, \mathbf{p}, t) \to f(\mathbf{r}, \mathbf{p}) + \delta f(\mathbf{r}, \mathbf{p})u(t)$, one denotes this the *steady-state* solution as well.

The BE is rather difficult to solve in certain cases, either because of the complexity of the non-linear scattering kernels or because of the involvement of several species that share the total available energy and momentum. An example of the latter is the electrical conduction in highly purified metals where electron-phonon scattering affords the principal mechanism of electrical resistance.

We note, in passing, the existence of a variational principle that allows a direct calculation of average transport parameters even if the BE proves intractable. In the case of electrical transport the variational integral is constructed in such a way that the exact (albeit, unknown) distribution yields an absolute minimum in the resistivity $1/\sigma$, with σ the electrical conductivity. Thus if the resistivity were calculated using a suitable "trial" distribution and minimized with respect to the adjustable parameters, the result would be a satisfactory upper bound to the exact answer, i.e. a lower bound to the exact σ. The derivation of this theorem found in John Ziman's book[g] is based on the original work of Lord Rayleigh[h] in the context of hydrodynamics, as generalized subsequently by Onsager;[i] it is called the *principle of least entropy production.*

7.8. Electrical Currents in a Low-Density Electron Gas

Initially let us assume a constant but small charge density allowing us to study the electron gas in the classical limit. Physically this scenario applies to homogeneous n-type semiconductors but, after a trivial change in the sign of the charge e, it applies to p-type semiconductors as well. After minor modification, it applies to metals as well. However, interfaces, such as found at rectifying junctions, create a density gradient that requires a special treatment beyond the scope of the present discussion.

In the absence of a density gradient there is no diffusion term. Assume a spatially homogeneous time-dependent electrical field $\mathbf{E}_\omega \cos \omega t$ along the z-axis. For charge carriers of charge e and effective mass m the BE takes on

[g]J. Ziman, *Electrons and Phonons*, Oxford, 1960, pp. 280 *ff.*
[h]Lord Rayleigh (J. W. Strutt), *Phil. Mag.* **26**, 776 (1913).
[i]L. Onsager, *Phys. Rev.* **37**, 405 and **38**, 2265 (1931), L. Onsager and S. Machlup, *Phys. Rev.* **91**, 1505 and 1512 (1953).

the following form:

$$\left[\frac{\partial}{\partial t} + eE_\omega \cos \omega t \frac{\partial}{\partial p_z}\right] f(p,t) = -C_1\{f(p,t)\} - C_2\{f(p,t)\}, \qquad (7.19)$$

where C_2 was defined in (7.17b) and C_1 is an additional, one-body collision operator, corresponding to *elastic* collisions of the individual charge carriers with impurity atoms, surfaces and other imperfections spoiling the translational invariance of the crystal. Explicitly,

$$C_1\{f(p,t)\} = \int d^3p' K(p|p')\delta(\varepsilon(p) - \varepsilon(p'))\{f(p,t) - f(p',t)\}. \qquad (7.20)$$

While this integral is on the "energy shell" it does not conserve momentum. Assuming the kernel K to be even in $\mathbf{p}$ and $\mathbf{p'}$,[j] this takes the generic form:

$$C_1\{f(\mathbf{p},t)\} = \frac{f(\mathbf{p},t) - \varphi(p,t)}{\tau(p)} \qquad (7.21)$$

where $\frac{1}{\tau(p)} \equiv \int d^3p' K(p|p')\delta(\varepsilon(p) - \varepsilon(p'))$ and $\varphi(p,t) \equiv \langle f(\mathbf{p},t)\rangle$ are both isotropic. The quantity $\tau(p)$ is the "mean-free-time" of a carrier of energy $\varepsilon(p)$ between "collisions" and is related to the "mean-free-path" $l(p)$ by $\tau(p) = l(p)/v(p)$, where $v(p) = \partial\varepsilon/\partial p = p/m$. In the event l is constant, $\tau(p) = ml/p$.

To illustrate in closed form the possibilities arising from the BE, let us choose $l(p) \propto p$ instead, hence $\tau(p) = \tau =$ constant. This results in a "toy model" that can be solved for the time dependence of the current $\mathbf{j}_z$ and of the Joule heating $\mathbf{j} \cdot \mathbf{E}$ without our ever solving for $f(\mathbf{p},t)$. The procedure is as follows:

Multiply both sides of the BE by $e\tau v_z/V$ and sum over all p. The term in C_2 vanishes by conservation of momentum, while — with C_1 in the form of Eq. (7.21) — one sees that the sum over $v_z\varphi(p,t)$ in C_1 vanishes by symmetry. What remains is,

$$\tau\frac{\partial j_z}{\partial t} + \frac{e^2}{m}\tau E_\omega \cos \omega t \frac{1}{V}\sum_p p_z\frac{\partial f(p,t)}{\partial p_z} = -j_z \qquad (7.22)$$

We change the sum into an integral and follow this by a partial integration on p_z. This results in an expression in which only $n \equiv N_{el}/V$ and $j = j_z$ figure:

$$\tau\frac{\partial j}{\partial t} - \frac{e^2\tau}{m}nE_\omega \cos \omega t = -j. \qquad (7.23)$$

[j]E.g. a function of p and p' only.

The quantity $\mu_n \equiv e^2\tau/m$ is defined as the electrons' *mobility*, a measure of the quality of the material. The *d-c* electrical conductivity $\sigma_o \equiv n\mu_n$. Assuming the electric field is "turned on" at $t = 0$, the explicit solution of this first-order *d.e.* for $j(t)$ is

$$j(t) = \frac{\sigma_o}{1 + (\omega\tau)^2} E_\omega\{\cos\omega t + (\omega\tau)\sin\omega t - e^{-t/\tau}\}. \qquad (7.24)$$

Ohm's law is satisfied: the current is linearly proportional to the electric field. The electrical *conductivity* is the in-phase component, $\sigma_R(\omega) = \sigma_o/[1+ (\omega\tau)^2]$. The out-of-phase or *inductive* component is $\sigma_I(\omega) = -\sigma_o\omega\tau/[1 + (\omega\tau)^2]$. Together they yield the well-known "Drude conductivity" of the classical electron gas.[k] The exponential represents the transient response to the step function at $t = 0$.

Next, multiply the BE by $p^2/2mV$ and sum over p. The collision terms vanish identically (both C_1 and C_2 are "elastic", i.e. conserve energy) and the remainder yields an equation for for the rate of increase of the total kinetic-energy-density $e(t) = E(t)/V$,

$$de(t)/dt = (E_\omega \cos\omega t)j(t) \qquad (7.25)$$

with $j(t)$ given in the preceding equation. When integrated, this gives the equation for the time-dependence of the "Joule heating" of the electron gas. Asymptotically, the energy shows a linear rate of increase, $e(t) \approx (t/2)\sigma(\omega)(E_\omega)^2$.

Problem 7.5.   Carry out the derivation of Eq. (7.25). Then perform the calculations outlined above to obtain the carriers' kinetic energy as a function of t. Distinguish the systematic increase, the out-of-phase oscillations and the transient contribution under the two different scenarios: $\tau\omega \gg 1$ and $\tau\omega \ll 1$ [*N.B.*: the *d-c* limit is $\omega \to 0$.] [*Hint*: express t in units of $\approx 2/\omega$.]

We conclude that in the absence of inelastic collisions there can *not* be a *true* asymptotic steady state! (For the energy stored in the electron gas increases without bound; if it were legitimate to assign a temperature to this ensemble, it too would increase linearly with t.)

---

[k]Had we used a complex driving field $E_\omega \exp(i\omega t)$, the complex conductivity resulting from this equation would be written as $\sigma_R(\omega) + i\sigma_I(\omega) = \sigma_o/[1 + i\omega\tau]$.

7.9.   Diffusion and the Einstein Relation

Integration of the BE over all p yields the equation of continuity,

$$\frac{\partial n(\mathbf{r}, t)}{\partial t} + \nabla \cdot \mathbf{j}(\mathbf{r}, t) = 0 \qquad (7.26)$$

with $n(\mathbf{r}, t)$ the local carrier density and $\mathbf{j}(\mathbf{r}, t)$ the local particle-current density. Note that, by detailed balance, both C_1 and C_2 vanish upon being summed and do not appear in this result.

Fick's Law expresses empirically a linear dependence of diffusion currents on the concentration gradient. With D the diffusion coefficient, Fick's law states:

$$\mathbf{j}_D(\mathbf{r}, t) = -D\nabla n(\mathbf{r}, t). \qquad (7.27)$$

Combining it with the equation of continuity yields the *diffusion equation*:

$$\frac{\partial n(\mathbf{r}, t)}{\partial t} - D\nabla^2 n(\mathbf{r}, t) = 0. \qquad (7.28)$$

Fourier transformation of the diffusion equation yields the signature *dispersion for diffusion*: $\omega = iDk^2$.

Now, in a charged gas consisting solely, say, of electrons, Poisson's equation relates $en(\mathbf{r}, t)$ to $\nabla \cdot E(\mathbf{r}, t)$. So it seems the diffusion equation can relate the conductivity to the diffusion coefficient. Actually, Einstein was first to consider the sum of charge currents, $\sigma_0 E - eD\nabla n$, with $\sigma_0 = n_0 \mu_n$; using the barometer equation, $n = n_0 \exp(-V(\mathbf{r})/kT)$ and $\mathbf{E} = -\nabla V$ he extracted the "Einstein relation" connecting the diffusion coeficient and the mobility:

$$D = \frac{kT}{|e|}\mu, \qquad (7.29)$$

upon setting $\mathbf{j}_{\text{tot}} = 0$ in thermodynamic equlibrium. This formula applies to electrons and/or holes separately in semiconductors and to other types of diffusing charge carriers, including vacancies and inerstitials in semiconductors and in ionic crystals.

7.10.   Electrical Conductivity of Metals

In a dense electron gas such as that in a metal, the streaming terms in the BE are formally unchanged, remaining precisely as given in Eq. (7.17a). Of course, the initial conditions are different. At $t = 0$ the initial distribution in the semiconductor is a Maxwell–Boltzmann distribution, while in the metal it is the Fermi–Dirac distribution at the initial time. But if σ can be obtained knowing just the particle density n — without even requiring details about f — such distinctions become irrelevant!

On the collision side of the BE, C_1 is unaltered, because in one-body scattering processes the factor $\{f(1-f')-f'(1-f)\}$ reduces to $\{f-f'\}$, hence the exclusion principle plays no apparent rôle. Also, in a dense, degenerate, electron gas only those states within kT of the Fermi surface are important. Thus, in metals, there is little error in setting $\tau(p) = \tau(p_F) = $ constant in Eq. (7.21). On the other hand C_2 is affected by the exclusion principle. It now takes the form:

$$
C_2\{f(\mathbf{r}, \mathbf{p}, t)\} = \iiint d^3p_2 d^3p'_1 d^3p'_2 K(p, p_2|p'_1, p'_2)
$$
$$
\times \delta(\mathbf{p} + \mathbf{p}_2 - \mathbf{p}_{1'} - \mathbf{p}_{2'})\delta(\varepsilon + \varepsilon_2 - \varepsilon_{1'} - \varepsilon_{2'})
$$
$$
\times \{f(p, t)f(p_2, t)(1 - f(p'_1, t))(1 - (p'_2, t))
$$
$$
- f(p'_1, t)f(p'_2, t)(1 - f(p, t))(1 - f(p_2, t))\}. \quad (7.30)
$$

But the procedure outlined in Eqs. (7.22)–(7.25) never involved C_2![1] Thus, the formulas obtained for the dilute gas remain true for metals. Explicit proof is given in the following Problem.

Problem 7.6.   Repeat the calculations in Eqs. (7.19)–(7.23) using C_2 appropriate to Fermi-Dirac statistics, as given above, to re-derive Eqs. (7.24)–(7.25). Compare magnitudes of τ and of n in the metal to those in the semiconductor, using reasonable estimates.

7.11.   Exactly Solved "Backscattering" Model

Let us revisit Eqs. (7.19)–(7.21), replacing the isotropic kernel by one sharply biased in the forward and backward directions, i.e. by $K \propto \delta(1 - \cos^2\theta_{p,p'})$, where $\theta_{p,p'} = \cos^{-1}\mathbf{p}\cdot\mathbf{p}'/pp'$ is the scattering angle. The following replaces Eq. (7.21):

$$
C_1(f(\mathbf{p}, t)) = \frac{f(\mathbf{p}, t) - f(-\mathbf{p}, t)}{2\tau}. \quad (7.31)
$$

---

[1]Because C_2 explicitly conserves total momentum, two-body collisions *can not* affect the total current in cases where the effective mass tensor can be defined, i.e. where $v_x \propto p_x$, $v_y \propto p_y$, etc., as is usually the case in semiconductors.

This defines the "backscattering model".[m] We found previously in Eqs. (7.22) *ff* that if τ is independent of p one can obtain the time dependence of the current density without any knowledge of the distribution function whatsoever. With the special scattering function defined here, this earlier cnclusion remains substantially unchanged, but it will now be possible to determine what kind of $f(\mathbf{p}, t)$ makes the steady-state current possible.

With the electric field along the z-direction it seems clear that the distribution in p_x and p_y is affected by neither the streaming terms nor by the collisions. For the low density gas it is thus possible to factor f as,

$$f(\mathbf{p}, t) \propto \exp\left(\frac{-p_x^2 - p_y^2}{2mkT}\right) f(p_z, t).$$

This reduces the BE to a one-dimensional p.d.e. equation for $f(p_z, t)$,

$$\left[\frac{\partial}{\partial t} + eE\frac{\partial}{\partial p_z}\right] f(p_z, t) = -\left[\frac{f(p_z, t) - f(-p_z, t)}{2\tau}\right]. \tag{7.32}$$

Although we only analyze the dilute case in this text, the diligent reader may wish to apply the same method to the metallic limit.[n]

First, introduce the even and odd parts of f in p_z, denoted F and G respectively, as the new dependent variables. Each satisfies a different equation:

$$\frac{\partial F(p_z, t)}{\partial t} + eE\frac{\partial G(p_z, t)}{\partial p_z} = 0 \tag{7.33a}$$

and

$$\frac{\partial G(p_z, t)}{\partial t} + eE\frac{\partial F(p_z, t)}{\partial p_z} = -\frac{G(p_z, t)}{\tau} \tag{7.33b}$$

Differentiation of (7.33a) *w.r.* to t and (7.33b) *w.r.* to p_z yields a second-order differential equation in F,

$$\frac{\partial^2 F}{\partial t^2} + \frac{1}{\tau}\frac{\partial F}{\partial t} - (eE)^2\frac{\partial^2 F}{\partial p_z^2} = 0. \tag{7.34}$$

The solution to this wave equation is easily expressed in the "natural" variables $t' = t/2\tau$, $x_0 = (2mkT)^{-1/2}$ and $\alpha(x) = [1 - (2\tau eEx)^2]^{1/2}$. It

---

[m]This section is extracted from D. C. Mattis, A. M. Szpilka and H. Chen, *Mod. Phys. Lett.* **B3**, 215 (1989); also, see J. Palmeri, *J. Stat. Phys.* **58**, 885 (1990), who extends the work and derives the connection to hydrodynamics.

[n]In the case of a dense electron gas (metal), f is not separable but takes the form $f(\varepsilon_{x,y}, p_z, t)$, with $\varepsilon_{x,y} = (p_x^2 + p_y^2)/2m$ remaining fixed. The initial condition is then f = Fermi–Dirac function at $t = 0$. The p.d.e. is then solved, exactly as here, although with this different initial condition it has a different outcome.

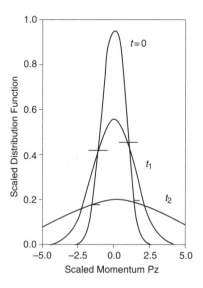

Fig. 7.2.   Evolution of distribution function with time.

Exact distribution function $f(p_z,t)$ as given in Eq. (7.35), at $t=0$, $t_1 = 500 \times (2\tau)$, and $t_2 = 5000 \times (2\tau)$. The scaled momentum is in units of $1/x_0$. The parameter $2\tau eEx_0$ is set $= 0.1$ in this calculation. Asymmetry in the distribution function $f(p_z,t)$ is highlighted by horizontal line segments. Despite all external appearances, f continues to evolve (broaden and flatten out in time), all the while the current density remain strictly constant and ohmic (proportional to the electric field).

is subject to initial conditions appropriate to a dilute-gas: at $t=0$, $f =$ Maxwell–Boltzmann distribution. Explicitly,

$$F(p_z,t) = \frac{1}{4\pi}e^{-t'}\int_{-\infty}^{\infty} dx (\cos p_z x)e^{-(\frac{x}{2x_0})^2}$$

$$\times \left[\left(1+\frac{1}{\alpha(x)}\right)e^{t'\alpha(x)} + \left(1-\frac{1}{\alpha(x)}\right)e^{-t'\alpha(x)}\right] \quad (7.35a)$$

The odd part can then be inferred from (7.33a),

$$G(p_z,t) = \frac{\tau eE}{2\pi}e^{-t'}\int_{-\infty}^{\infty} dx\frac{x\sin p_z x}{\alpha(x)}e^{-(\frac{x}{2x_0})^2}$$

$$\times [e^{t'\alpha(x)} - e^{-t'\alpha(x)}] \quad (7.35b)$$

The desired result $f(p_z,t) = F(p_z,t) + G(p_z,t)$ is plotted in Fig. 7.2 as a function of p_z at various values of t.

Conclusion: despite the absence of a mechanism for heat dissipation it is possible for the current to evolve into an *ohmic* steady state. However, absent any plausible mechanisms for heat dissipation, the distribution function f continues to evolve forever. The introduction of any such additional mechanism should affect this conclusion mightily.

7.12. Electron-Phonon Scattering

The electron-phonon scattering mechanism allows both the momentum and the excess energy in the electron gas to be transfered to the phonons. Because the latter are neutral this process detracts from the overall electrical current.

The collision integral that we shall wish to add to the *rhs* of the BE in (7.19) is $- C_{e\text{-}ph}\{f(p,t), n_q(t)\}$. *In semiconductors* the equilibrium distributions, ensuring that $C_{e\text{-}ph} = 0$ initially, are the Maxwell-Boltzmann for f and the Planck for n_q. At later times, $t > 0$,

$$
\begin{aligned}
C_{e\text{-}ph}\{f(p,t), n_q(t)\} = \int d^3q K(q)\{&f(p,t)[(1 + n_{-q}(t))\delta(\varepsilon_p - \varepsilon_{p+q} - \hbar\omega_q) \\
&+ n_q(t)\delta(\varepsilon_p - \varepsilon_{p+q} + \hbar\omega_q)] \\
&- f(p+q,t)[(1 + n_q(t))\delta(\varepsilon_p - \varepsilon_{p+q} + \hbar\omega_q) \\
&+ n_{-q}(t)\delta(\varepsilon_p - \varepsilon_{p+q} - \hbar\omega_q)]\}
\end{aligned}
\tag{7.36}
$$

in which both ω_q and $K(q)$ are assumed to be even functions of $\mathbf{q}$.[o] Clearly there is no qualitative change in the solution to the electrons' BE *except that* τ *decreases* with an increasing number of phonons. In a closed system, a steadily increasing number of phonons picks up momentum and energy from the electrons at a rate determined by a collision integral rather similar to the above. We should also write a BE for the phonons. Coupled equations of this type have recently been solved to a good approximation so as to yield details of the decay of nonequilibrium populations of carriers and/or phonons.[P]

But at this juncture the reader might be curious to learn of a simpler and general procedure, that allows to deal with all manners of transport and scattering mechanisms.

---

[o]In metals, $f(p,t)$ in the above is replaced by $f(p,t)[1 - f(p+q,t)]$ and $f(p+q,t)$ by $f(p+q,t)[1 - f(p,t)]$. In metals momentum non-conservation, associated with Bragg scattering (*umklapp*) of carriers at the FS, i.e. those with relatively high momentum, plays an essential rôle. For details see J. Ziman, *op. cit.*[g]
[P]See P. B. Allen, *Phys. Rev. Lett.* **59**, 1460 (1987).

7.13. Approximating the Boltzmann Equation

In the preceding section we briefly examined a nonlinear scattering term and had to be satisfied with just an estimate of the behavior (although quantitative results do exist in the literature.[g,q]) In a metal the terms inside the integral appear to be even more highly nonlinear owing to the presence of the extra Pauli-principle factors of $(1-f)$. If, however, we had determined that the forces and/or the spatial gradients were especially weak compared to kT (or to ε_F in a metal), and that a steady-state close to thermodynamic equilibrium had been established, a set of simplifications impose themselves.

For example, assume that $f(p,t)$ differs from its thermal average $f_0(\varepsilon/kT)$ and $n_q(t)$ from the corresponding Planck function $n_0(q) = n(\hbar sq/kT)$ by small quantities only, linear in the perturbing forces or density gradients. After we discard terms in the BE that are nonlinear in the perturbing forces or density gradients, what remains can, as in most simple linear equations, be solved explicitly. This approximation is called "linearizing the BE".

A second approximation, generally denoted the "relaxation-time approximation" to the BE, is more severe and, as we shall determine subsequently, harder to justify. Let us illustrate it with the aid of the one-body collision term C_1 of Eq. (7.21), which is already inherently linear in f.

The relaxation-time approximation consists of replacing $\varphi(p,t)$ by $f_0(\varepsilon/kT)$, ensuring that, asymptotically, $f \to f_0$ once the streaming term is turned off. This approximation is most commonly used in the (otherwise unwieldy) calculation of transport coefficients, including electrical and thermal conductivities, and in the calculation of Hall coefficients (electrical transport in an external magnetic field).[q]

In semiconductors, where the dispersion relation of conduction particles differs significantly from the simple $\varepsilon = p^2/2m$, transport coefficients have to be expressed as components of a tensor. The relaxation-time approximation provides a simple way to calculate them. Let us illustrate this, first by solving for the electrical conductivity of a Maxwell–Boltzmann gas of charge carriers, of arbitrary dispersion $\varepsilon(p)$ and scattering lifetime, $\tau_p \equiv \tau(\varepsilon(p))$. With static electric field E now assumed to be along the x-direction, the linearized BE is just,

$$eEv_x\frac{\partial f_0(\varepsilon_p)}{\partial \varepsilon_p} = -\frac{f(p) - f_0(\varepsilon_p)}{\tau_p} \tag{7.37}$$

---

[q]These and many other examples of its use are calculated in A. Smith, J. Janak and R. Adler, *Electronic Conduction in Solids*, McGraw-Hill, New York, 1967.

where $v_x \equiv \partial\varepsilon_p/\partial p_x$. The error is $O(E \times (f - f_0))$, i.e. it is quadratic in the external field parameter. Equation (7.37) is solved by inspection: $f = f_0[1 + h]$, where $h_p = eE\tau_p v_x/kT$ is linear in the small quantity E. Hence the electrical current density along the α axis is

$$j_\alpha = \frac{e^2 E}{kT} \int d^3p f_0(\varepsilon_p)\tau_p v_x v_\alpha$$

(where $\alpha = x, y$, or z) and the components of the conductivity tensor defined by $j_\alpha = \sum_\beta \sigma_{\alpha,\beta} E_\beta$ are,

$$\sigma_{\alpha,\beta} = \frac{e^2}{kT} \int d^3p f_0(\varepsilon_p)\tau_p v_\alpha v_\beta, \tag{7.38a}$$

with f_0 normalized to the particle density, $\int d^3p f_0(\varepsilon_p) = n_{el}$. This yields,

$$\sigma_{\alpha,\beta} = \frac{n_{el}e^2}{kT} \langle \tau_p v_\alpha v_\beta \rangle_{TA} . \tag{7.38b}$$

For spherical energy surfaces, $\varepsilon_p = m\mathbf{v}^2/2$, (7.38b) simplifies further, to $\sigma_{\alpha,\beta} = \sigma_0 \delta_{\alpha,\beta}$ with $\sigma_0 = n_{el}e^2\tau/m$. Here $\tau \equiv \langle 2\tau_p \varepsilon_p/3kT \rangle_{TA}$ is a lumped parameter.

The linearized approximation $f = f_0[1 + h]$ fails if $|h|$ is not $\ll 1$ as originally postulated. Where this postulate fails (i.e. at large $|v_x| \propto 1/|E|$) it is the pre-factor that becomes vanishingly small, $f_0 \propto \exp(\frac{-mkT}{2|eE\tau_p|^2})$. Thus, although it is not exact, the linearized solution is seen to be accurate *where it counts*.

7.14.   Crossed Electric and Magnetic Fields

The electric field is not the only force that can be exerted on a charge carrier. In crossed electric and magnetic fields, the Lorentz force is responsible for what is in effect an induced, transverse, electric field, first discovered in 1879 by E. H. Hall. With the applied electric field along the x-direction and the current flowing along the same direction, assume the magnetic field B to be aligned along the z-axis of a rectangular bar. The Lorentz force is along the y-axis; but as no current flows in this direction it must be countered by an equal and opposite electric force E_y caused by an accumulation of charge carriers. E_y is proportional to j_x and to B. The constant of proportionality is the Hall constant, R_H, and the non-diagonal component $\sigma_{x,y}$ is the *Hall conductivity*.

A simple calculation yields $R_H = 1/nec$, where $n =$ density of carriers and e is their charge.[r] The Hall conductivity is $\sigma_{x,y} = nec/B$. The arguments are as follows: for no *net* force along the y-axis, $E_y = \langle v_x \rangle B/c$. Then set $\langle v_x \rangle = j_x/ne$ to obtain the stated result forthwith. The generalization of Eq. (7.37) is,

$$e \left[E_x \frac{\partial f_0}{\partial p_x} + E_y \frac{\partial f_0}{\partial p_y} \right] + \frac{eB}{c} \left[v_y \frac{\partial f}{\partial p_x} - v_x \frac{\partial f}{\partial p_y} \right] + \mathbf{v} \cdot \frac{\partial f}{\partial \mathbf{r}}$$

$$= -\frac{f - f_0(\varepsilon_p)}{\tau_p}. \tag{7.39}$$

If one specializes to $\varepsilon_p = \mathbf{p}^2/2m = m\mathbf{v}^2/2$ and generalizes to time-dependent applied and induced electric fields, the substitution $f = f_0[1 + h]$, followed by division by f_0, yields:

$$-e \left[E_x(t) \frac{p_x}{m} + E_y(t) \frac{p_y}{m} \right] \frac{\partial \log f_0(\varepsilon_p)}{\partial \varepsilon_p}$$

$$= \frac{eB}{mc} \left[p_y \frac{\partial h}{\partial p_x} - p_x \frac{\partial h}{\partial p_y} \right] + \mathbf{v} \cdot \frac{\partial h}{\partial \mathbf{r}} + \frac{\partial h}{\partial t} + \frac{h}{\tau_p} \tag{7.40}$$

a *p.d.e.* in p_x, p_y, $\mathbf{r}$ and t for $h(\mathbf{r}, \mathbf{p}, t)$. The quantity $|eB/mc| \equiv \omega_c$ is the "cyclotron frequency" for carriers of charge e and mass m. This equation can be solved by the usual methods, that is: to any special solution add the general solution of the *rhs* (homogeneous in h) chosen so as to satisfy any imposed boundary conditions in space and time. The first such calculations were carried out by Fuchs and others half a century ago, in an effort to gauge the effects of surfaces on electrical conductivity in thin metallic wires and in thin films, and later in the calculation of the Hall efect and cyclotron resonance in semiconductors and in metals.[s]

In the case of the *dc* Hall effect, Eq. (7.39) suffices if we can approximate τ_p by a constant, τ, as is demonstrated in the following Problem.

---

[r]That is, R_H is negative for electrons and positive for holes.

[s]K. Fuchs, *Proc. Camb. Phi. Soc.* **34**, 100 (1938). E. H. Sondheimer, *Advances in Phys.* **1**, 1 (1952) and *Proc. Roy. Soc.* **A224**, 260 (1954). For conductivity and Hall effect with anisotropic energy surfaces in semiconductor thin films: F. S. Ham and D. C. Mattis, *IBM Journal* **4**, 143 (1960).

Problem 7.7.   Assuming τ_p is independent of p and the density n is consant, multiply both sides of Eq. (7.39) by $\tau e v_x$ and sum over all $\mathbf{p}$; repeat with $\tau e v_y$. Using the boundary condition $j_y = 0$ to determine E_y, obtain R_H and $\sigma_{x,y}$ without actually calculating f.

7.15.   Propagation of Sound Waves in Fluids

We know from common experience that sound waves propagate at a speed s after being generated, while their intensity decays exponentially in a time τ. Experiment has shown that both these quantities are functions of the wavelength. The following theory of sound propagation confirms this. It is based on a linearization of the BE and is appropriate to small amplitude waves and therefore not to large amplitude noises or "shock waves".

For molecules in a dilute gas only two-body collisions enter into Eq. (7.18). They provide both the mechanism by which sound is propagated and that by which it is dissipated (as the energy and momentum in the collective mode becomes transferred to the incoherent motion of individual molecules). Again we write $f = f_0(1 + h)$ and retain only term linear in h. Thereupon Eq. (7.17b) becomes:

$$C_2(f(p, r, t)) = f_0(p) \int d^3p_2 \int d^3p_1{}' \int d^3p_2{}' K(p, p_2 | p_1{}', p_2{}')$$

$$\times \delta(p + p_2 - p_1{}' - p_2{}')\delta(p^2 + p_2{}^2 - p_1{}'^2 - p_2{}'^2)f_0(p_2)$$

$$\times \{h(p, r, t) + h(p_2, r, t) - h(p_1{}', r, t) - h(p_2{}', r, t)\} \quad (7.41)$$

assuming the collision occurs over a small distance so $r_2 \approx r$. It is convenient to rewrite the above as $C_2(f) \equiv n f_0(p) C(h(p, r, t))$, where $\mathbf{C}$ is a linear, integral, operator on h, having a complete spectrum of eigenfunctions Ψ_j and eigenvalues $\lambda_j \equiv 1/2\tau_j$ (see footnote v, *infra*) and $n = N/N_0$ is the dimensionless particle density N/L^3 in units of a reference density N_0/L^3. We shall use a number of theorems taken from linear algebra. Consider, for example,

$$\mathbf{C}(\Psi_j) = \lambda_j \Psi_j . \quad (7.42)$$

The eigenvalues λ and eigenfunctions Ψ of the 2-body collision operator are infinite in number and are known explicitly only in special instances.[t]

---

[t] L. Waldmann, *Handbuch d. Physik* Vol. 12.

They usually have to be obtained by numerical means. As we shall show
shortly, all λ's are non-negative; the lowest 5 eigenvalues are zero while all the
others are positive. Unless the scattering kernel is singular or nonintegrable,
they are also bounded from above, lying in the interval $\lambda_{min} < \lambda < \lambda_{max}$.
In any case, it is always permissible to expand the scattering operator in
its eigenfunctions and eigenvalues, as we shall do shortly. We express the
orthonormality of the eigenstates (with f_0 a weighting function) through

$$\int d^3p f_0(p)\Psi_\alpha^+(p)\Psi_\beta(p) = \delta_{\alpha,\beta}. \tag{7.43}$$

Having factored out n, henceforth we normalize $f_0(p)$ to unity: $\int d^3p f_0(p) = 1$. The *completeness* of the Ψ's means that an *arbitrary* function $h(p)$ can be
expanded in a "Fourier" series

$$h(p) = \sum_\alpha A_\alpha \Psi_\alpha(p), \text{ with } A_\alpha = \int d^3p' f_0(p')\Psi_\alpha^+(p')h(p'). \tag{7.44}$$

Inserting this expression for A_α into the expansion of h yields,

$$h(p) = \sum_{\text{all } \alpha} \left[\int d^3p' f_0(p')\Psi_\alpha^+(p')h(p')\right]\Psi_\alpha(p)$$

$$= \int d^3p' \left\{ f_0(p') \sum_{\text{all } \alpha} \Psi_\alpha^+(p')\Psi_\alpha(p) \right\} h(p').$$

As h is arbitrary this uniquely determines the $\{\cdots\}$ above,

$$\delta(p-p') = f_0(p') \sum_{\text{all } \alpha} \Psi_\alpha^+(p')\Psi_\alpha(p)$$

$$= f_0(p') \sum_{\text{all } \alpha} \Psi_\alpha^+(p)\Psi_\alpha(p'). \tag{7.45}$$

Problem 7.8. Prove that the preceding Fourier expansion of h is essentially exact by showing *explicitly*:

$$\int d^3p f_0(p) \left| h(p) - \sum_\alpha A_\alpha \Psi_\alpha(p) \right|^2 = 0.$$

The scattering kernel has the following expansion in terms of its eigen-functions and eigenvalues:

$$C = f_0(p') \sum_{\gamma>5} \lambda_\gamma \Psi_\gamma^+(p') \Psi_\gamma(p) = f_0(p') \sum_{\gamma>5} \lambda_\gamma \Psi_\gamma^+(p) \Psi_\gamma(p'). \qquad (7.46)$$

In this form it automatically satisfies the eigenvalue equation, (7.42):

$$C(\Psi_\alpha) = \int d^3p' \left[f_0(p') \sum_{\gamma>5} \lambda_\gamma \Psi_\gamma^+(p') \Psi_\gamma(p) \right] \Psi_\alpha(p')$$

$$= \lambda_\alpha \Psi_\alpha(p). \qquad (7.47)$$

As we shall show, the first 5 eigenfunctions correspond to the 5 constants of the motion and have $\lambda = 0$. Therefore they need not be included in the sums above. Upon multiplication of both sides of the equation above by $f_0(p)\Psi_\beta^+(p)$ and integration over all p, one obtains:

$$\int d^3p f_0(p) \Psi_\beta^+(p) C(\Psi_\alpha) = \lambda_\alpha \delta_{\beta,\alpha}, \qquad (7.48)$$

proving that the collision operator is a diagonal operator ($\alpha = \beta$) in the representation of its eigenfunctions, the Ψ's.

Now, the *lhs* of Eq. (7.47) originally stood for

$$C(\Psi_\alpha) = \int d^9p K(p, p_2|p_1', p_2')$$

$$\times \delta(p + p_2 - p_1' - p_2')\delta(p^2 + p_2^2 - p_1'^2 - p_2'^2) f_0(p_2)$$

$$\times \{\Psi_\alpha(p) + \Psi_\alpha(p_2) - \Psi_\alpha(p_1') - \Psi_\alpha(p_2')\}.$$

Inserting this definition into the *lhs* of Eq. (7.48) with $\beta = \alpha$, we obtain:

$$\int d^3p f_0(p) \Psi_\alpha^+(p) \left[\int d^9p K(p, p_2|p_1', p_2')\delta(p + p_2 - p_1' - p_2') \right.$$

$$\left. \times \delta(p^2 + p_2^2 - p_1'^2 - p_2'^2) f_0(p_2)\{\Psi_\alpha(p) + \Psi_\alpha(p_2) - \Psi_\alpha(p_1') - \Psi_\alpha(p_2')\} \right]$$

$$= \frac{1}{4} \int d^{12}p K(p, p_2|p_1', p_2')\delta(p + p_2 - p_1' - p_2')\delta(p^2 + p_2^2 - p_1'^2 - p_2'^2)$$

$$\times f_0(p) f_0(p_2) |\{\Psi_\alpha(p) + \Psi_\alpha(p_2) - \Psi_\alpha(p_1') - \Psi_\alpha(p_2')\}|^2 \geq 0$$

$$= \lambda_\alpha. \qquad (7.49)$$

(Recall the similar construction in the H-Theorem.) The integrand is non-negative. There exist 5 linearly independent eigenstates having zero eigenvalue: the 5 constants of the motion, $\Psi = 1, p_x, p_y, p_z \& p^2$. All others, starting with Ψ_6, have eigenvalues $\lambda > 0$.

To determine the collective modes in the BE suppose the driving force in $\mathbf{F}(r,t) = (0,0,F_z(r,t))$ takes the form $F_z(r,t) = \sum_k e^{i(kr-\omega t)} F_{k,\omega}$. Within the linear approximation and according to the *superposition principle*, it is permissible to consider the response to each $F_{k,\omega}$ separately, as the response is linearly additive. Then the BE factors into separate equation for each mode. Each takes the form:

$$i\left(-\omega + \frac{1}{m}\mathbf{p}\cdot\mathbf{k}\right)h_{k,\omega}(p) + F_{k,\omega}\frac{\partial}{\partial p_z}\log f_0(\varepsilon_p) = -nC(h_{k,\omega}(p)). \quad (7.50)$$

The solution of Eq. (7.50) determines the response, *viz.*, $h_{k,\omega}(p) = F_{k,\omega}G_{k,\omega}$ in the Fourier decomposition of $h(r,t;p) = \sum_{k,\omega} h_{k,\omega}(p)e^{i(k\cdot r-\omega t)}$. A *resonance* occurs wherever $G_{k,\omega}$ diverges, e.g. at a pole $\omega_R(\mathbf{k}) - i\omega_I(\mathbf{k})$ in the complex plane. Fluids are isotropic. Thus the *orientation* of $\mathbf{k}$ is immaterial and the pole is at $\omega(k) = \omega_R(k) - i\omega_I(k)$, where $k = |\mathbf{k}|$. We identify the pole in the response function as a sound wave and the dependence of ω_R on k as its dispersion, with $\omega_I(k)$ measuring the rate of decay or half-inverse lifetime, $1/2\tau(k)$.[u] The study of dispersion involves only the homogeneous part of the BE. We now derive the coupled equations for a new eigenvalue problem that will determine $\omega = \omega_R(k) - i\omega_I(k)$.

$$\int d^3 p f_0(p)\Psi_\beta^+(p)i\left(-\omega + \frac{1}{m}\mathbf{p}\cdot\mathbf{k}\right)\sum_\alpha A_\alpha\Psi_\alpha(p)$$

$$= -n\int d^3 p f_0(p)\Psi_\beta^+(p)\sum_\alpha A_\alpha\lambda_\alpha\Psi_\alpha(p),$$

or:

$$\omega A_\beta - \frac{1}{m}\sum_\alpha A_\alpha\int d^3 p f_0(p)\Psi_\beta^+(p)\mathbf{k}\cdot\mathbf{p}\Psi_\alpha(p) = -in\lambda_\beta A_\beta. \quad (7.51)$$

These coupled equations have a solution *iff* a secular determinant vanishes, i.e. $\det\|M_{\beta,\alpha}\| = 0$, where .

$$M_{\beta,\alpha} = (\omega + in\lambda_\beta)\delta_{\beta,\alpha} - \frac{1}{m}\int d^3 p f_0(p)\Psi_\beta^+(p)\mathbf{k}\cdot\mathbf{p}\Psi_\alpha(p). \quad (7.52a)$$

---

[u]If the *energy* in the sound wave decays at a rate $\propto 1/\tau$, the rate of decay of the amplitude is, by definition, $1/2\tau$.

Let us work out the first 5 solutions, properly normalized and orthogonalized. Define $\mathbf{u} \equiv \mathbf{p}(2mkT)^{-1/2}$, $v \equiv \omega(m/2kT)^{1/2}$ and $\Gamma_\alpha \equiv n\lambda_\alpha(m/2kT)^{1/2}$. In these units $f_0(u) = \pi^{-3/2}\exp{-u^2}$ and $\Gamma_1, \ldots, \Gamma_5 = 0$ and the entries in the secular determinant are given by

$$M_{\beta,\alpha} = (v + i\Gamma_\beta)\delta_{\beta,\alpha} - \int d^3u f_0(u)\Psi_\beta^+(u)k \cdot u\Psi_\alpha(u). \qquad (7.52b)$$

Problem 7.9.   Show that $\Psi_1, \ldots, \Psi_5$ below are an orthonormal set in u, i.e. $\int d^3u f_0(u)\Psi_i^+(u)\Psi_j(u) = \delta_{i,j}$ with $f_0(u) = \pi^{-3/2}\exp{-u^2}$:

$$\Psi_1 = 1, (\Psi_2, \Psi_3, \Psi_4) = \sqrt{2}\mathbf{u}, \ \Psi_5 = (2/3)^{1/2}(\mathbf{u}^2 - 3/2). \qquad (7.53)$$

An elementary theory can be constructed using just the first 5 eigenfunctions. Choose the direction of $\mathbf{k}$ as the z-axis. The 5×5 secular determinant contains only the following nonzero entries:

$M_{\alpha,\alpha}=v$ for $\alpha=1, 2, 3, 4, 5$, $M_{1,4}=M_{4,1}= -k/\sqrt{2}$ and $M_{4,5}=M_{5,4} = -k/\sqrt{3}$.

Elementary algebra shows that three roots are zero, with the remaining two being $v = \pm k\sqrt{5/6}$. In physical units,

$$\omega = \pm k(5kT/3m)^{1/2}. \qquad (7.54)$$

This result is also well known from elementary thermodynamical considerations in the limit $k \to 0$ although clearly, the ideal gas theory does not deal adequately with the collisions and neither does the 5×5 matrix used here. To involve collisions, one must extend the theory and calculate the complex ω as a function of k using 6 or more eigenfunctions. Both real and imaginary parts of ω now involve the Γ's. A simple-minded extension is suggested in the following problem but it does not go far enough.

Problem 7.10.   Construct a sixth eigenstate extending the preceding results, by postulating $\Psi_6 = a((\mathbf{u}^2)^2+b\mathbf{u}^2+c)$, obtaining a, b, c by requiring Ψ_6 to be orthogonal to Ψ_1 and Ψ_5 and to be normalized. Calculate the nontrivial matrix element that connects Ψ_6 to Ψ_5. Using (7.52b), solve for the now complex $v(k)$, written in the form $v_R = \pm k\varphi(k/\Gamma_6)$, and $v_I = \Gamma_6\chi(k/\Gamma_6)$ and compare the functions φ and χ you have calculated with the results found in Sec. 7.16.

7.16.   The Calculations and Their Result

Although one might plausibly believe that by extending the set of basis
functions in which to construct the secular determinant one achieves higher
accuracy, the opposite may be true. It is known that the convergence is
slow and uncertain. A non-perturbational approach in which *all* the Ψ's are
included imposes itself, as outlined briefly below.

Rather than excluding an infinite number of states, keep the 5 eigenstates
with zero λ (the constants of the motion) and approximate the remaining
λ_j's $(j > 5)$ by λ_{min},[v] i.e.,

$$C(\Psi_\alpha) = \int d^3p' \left[f_0(p') \sum_{\gamma>5} \lambda_{min}\Psi_\gamma^+(p')\Psi_\gamma(p) \right] \Psi_\alpha(p')$$

$$= \begin{cases} \lambda_{min}\Psi_\alpha(p) \text{ for } \alpha > 5 \\ 0 \quad \text{ for } \alpha \leq 5 \end{cases} \tag{7.55a}$$

Let us invoke *completeness* to render this expression more transparent:

$$C(\Psi_\alpha) = \lambda_{min} \int d^3p'\delta(p-p')\Psi_\alpha(p')$$

$$- \lambda_{min} \int d^3p' \left[f_0(p') \sum_{\beta=1}^{5} \Psi_\beta^+(p')\Psi_\beta(p) \right] \Psi_\alpha(p')$$

$$= \lambda_{min}\Psi_\alpha(p) - \lambda_{min} \int d^3p' \left[f_0(p') \sum_{\beta=1}^{5} \Psi_\beta^+(p')\Psi_\beta(p) \right] \Psi_\alpha(p') \tag{7.55b}$$

We now include this formula in the homogeneous BE — including streaming
parts — as reëxpressed in the more convenient, dimensionless, language of
the $\mathbf{u}$'s, v's and Γ's.

The sound wave amplitude $\Phi(u)$ satisfies the integral equation:

$$[v+i\Gamma_{min}-ku_z]\Phi(u) = i\Gamma_{min}\pi^{-3/2} \int d^3u' e^{-u'^2}$$

$$\times \left[1 + 2\mathbf{u}\cdot\mathbf{u}' + \frac{2}{3}\left(u^2-\frac{3}{2}\right)\left(u'^2-\frac{3}{2}\right) \right] \Phi(u') \tag{7.56}$$

[v]It is even possible to verify the accuracy, by repeating the calculation using λ_{max}. In
cases where the spectrum between λ_{min} and λ_{max} is rather narrow the results of the two
calculations would nearly coïncide.

obtained by explicitly writing out the bilinear sum over the first 5 eigenfunctions given in Eq. (7.53). The kernel in the integral equation is the sum of 5 separable kernels. The equation itself is solved by defining a, $\mathbf{b}$ and c as follows:

$$a = \frac{1}{\pi^{3/2}} \int d^3 u' e^{-u'^2} \Phi(u'), \quad \mathbf{b} = \frac{1}{\pi^{3/2}} \int d^3 u' e^{-u'^2} \mathbf{u}' \Phi(u')$$

and

$$c = \frac{1}{\pi^{3/2}} \int d^3 u' e^{-u'^2} \left(u'^2 - \frac{3}{2} \right) \Phi(u') \tag{7.57}$$

Then Eq. (7.56) yields the explicit solution for Φ:

$$\Phi(u) = i\Gamma_{\min} \frac{[a + 2\mathbf{b} \cdot \mathbf{u} + \frac{2}{3}c(u^2 - \frac{3}{2})]}{[v + i\Gamma_{\min} - ku_z]}. \tag{7.58}$$

Inserting this into the integrals in (7.57) yields 5 equations in a, $\mathbf{b}$ and c. With each being explicitly complex, this substitution actually yields 10 linear equations in 10 unknowns. However, after making the self-consistent assumption $b_z \neq 0$, $b_x = b_y = 0$ their number is reduced to 3 complex equations in 3 complex unknowns; a, b_z, and c.

$$a = i\Gamma_{\min} \frac{2\pi}{\pi^{3/2}} \int_0^\infty du' u'^2 e^{-u'^2} \int_0^\pi d\vartheta' \sin \vartheta'$$

$$\times \frac{[a + 2b_z u' \cos \vartheta' + \frac{2}{3}c(u'^2 - \frac{3}{2})]}{[v + i\Gamma_{\min} - ku' \cos \vartheta']} \tag{7.59a}$$

$$b_z = i\Gamma_{\min} \frac{2\pi}{\pi^{3/2}} \int_0^\infty du' u'^2 e^{-u'^2} \int_0^\pi d\vartheta' \sin \vartheta'$$

$$\times u' \cos \vartheta' \frac{[a + 2b_z u' \cos \vartheta' + \frac{2}{3}c(u'^2 - \frac{3}{2})]}{[v + i\Gamma_{\min} - ku' \cos \vartheta']} \tag{7.59b}$$

and

$$c = i\Gamma_{\min} \frac{2\pi}{\pi^{3/2}} \int_0^\infty du' u'^2 e^{-u'^2} \int_0^\pi d\vartheta' \sin \vartheta'$$

$$\times \left(u'^2 - \frac{3}{2} \right) \frac{[a + 2b_z u' \cos \vartheta' + \frac{2}{3}c(u'^2 - \frac{3}{2})]}{[v + i\Gamma_{\min} - ku' \cos \vartheta']} \tag{7.59c}$$

These have no solution unless the determinant of the coefficients of a, b_z and c vanishes, that is:

$$\det \begin{Vmatrix} A_{11} A_{12} A_{13} \\ A_{21} A_{22} A_{23} \\ A_{31} A_{32} A_{33} \end{Vmatrix} = 0 \tag{7.60}$$

where

$$A_{11} = 1 - i\Gamma_{\min} J(1), \quad A_{12} = -i\Gamma_{\min} J(2u' \cos \vartheta'), \quad A_{21} = \frac{1}{2} A_{12},$$

$$A_{13} = -i\Gamma_{\min} J\left(\frac{2}{3} u'^2 - 1\right), \quad A_{31} = \frac{3}{2} A_{13},$$

$$A_{22} = 1 - i\Gamma_{\min} J(2u'^2 \cos^2 \vartheta'),$$

$$A_{23} = -i\Gamma_{\min} J\left(u' \cos \vartheta' \left(\frac{2}{3} u'^2 - 1\right)\right),$$

$$A_{32} = 3A_{23}, \quad A_{33} = 1 - i\Gamma_{\min} J\left(\frac{2}{3}\left(u'^2 - \frac{3}{2}\right)^2\right)$$

and

$$J(g) \equiv \frac{2\pi}{\pi^{3/2}} \int du' u'^2 e^{-u'^2} \int_0^\pi d\vartheta' \sin \vartheta' \frac{g}{[v + i\Gamma_{\min} - ku' \cos \vartheta']},$$

where g is any of the polynomials in u' and $\cos \vartheta'$ in the preceding list. The results simplify if we first perform the integration over ϑ'.

$$\int_0^\pi d\vartheta' \sin \vartheta' \frac{1}{[v + i\Gamma_{\min} - ku' \cos \vartheta']} = \frac{1}{ku'} \log \frac{v + i\Gamma_{\min} + ku'}{v + i\Gamma_{\min} - ku'},$$

$$\int_0^\pi d\vartheta' \sin \vartheta' \frac{\cos \vartheta'}{[v + i\Gamma_{\min} - ku' \cos \vartheta']} = \frac{v + i\Gamma_{\min}}{(ku')^2} \log \frac{v + i\Gamma_{\min} + ku'}{v + i\Gamma_{\min} - ku'} - \frac{2}{ku'},$$

and

$$\int_0^' d\vartheta' \sin \vartheta' \frac{\cos^2 \vartheta'}{[v + i\Gamma_{\min} - ku' \cos \vartheta']}$$

$$= \frac{(v + i\Gamma_{\min})^2}{(ku')^3} \log \frac{v + i\Gamma_{\min} + ku'}{v + i\Gamma_{\min} - ku'} - 2\frac{v + i\Gamma_{\min}}{(ku')^2}$$

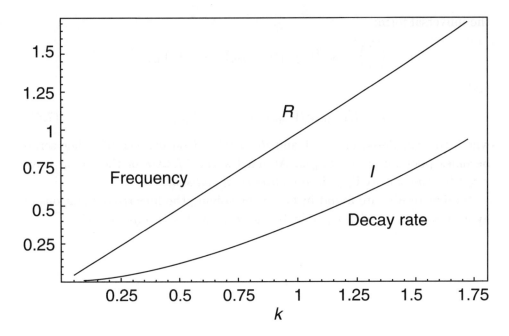

Fig. 7.3.   Dispersion and decay of sound waves.[w]

Universal plots based on Eq. (7.60). *Vertical axis* (in units of $\Gamma_{\min} \equiv$ the effective two-particle scattering cross-section times the density): the frequency v_R (curve labeled R) and decay rate v_I (labeled I) of a sound wave.

Horizontal axis: the wavevector k (also in units of $\Gamma_{\min}$). The curves are, to good accuracy, linear and parabolic respectively. Within the approximations of this theory, the speed of sound $s = v_R/k$ remains accurately constant over the entire range of wavelengths where the sound wave can propagate, i.e. for k not exceeding $O(1.75\Gamma_{\min})$ (wavelengths λ no shorter than $\approx \frac{\pi}{2}\Gamma_{\min}$). This illustrates an altogether non-intuitive but inescapable conclusion: the denser the gas, i.e. the stronger the collision parameter γ, the better the sound propagation!

The integrations over u' in the J's have to be performed numerically. The corresponding $A_{n,m}$ are inserted into (7.60) to find the nontrivial roots $v_R - iv_I$ as functions of k and $\Gamma_{\min}$. With the results displayed on the next page, we note from an expansion in powers of k that the solutions have to be in

[w]These results and the graphs in Fig. (7.3) are based on the numerical evaluation of Eq. (7.60) obtained by Dr. Prabasaj Paul at the University of Utah using *Mathematica*. The author is grateful to Dr. Paul for his help.

the universal form:

$$v_R = \left(\frac{5}{6}\right)^{1/2} k \cdot [1 + c_1 (k/\Gamma_{min})^2 + c_2 (k/\Gamma_{min})^4 + \cdots]$$

and

$$v_I = d_1 \Gamma_{min} (k/\Gamma_{min})^2 [1 - d_2 (k/\Gamma_{min})^2 + \cdots] \qquad (7.61)$$

where the calculated $c_1 \approx 0.1$ and $d_1 \approx 0.5$.[w] Convergence of either series becomes poor for $k > 1.5\Gamma_{min}$. At fixed k the *lifetime* of the sound wave $\tau \propto 1/v_I$ *vanishes* $\propto \Gamma_{min}$ *in the absence of collisons*.

Similar results are found in *liquids* by solving the linearized equations of hydrodynamics — with $w_I \propto k^2$ being associated with *diffusivity*.[x]

---

[x]N.H. March and M. P. Tosi, *Introduction to Liquid State Physics*, World Scientific, 2002; p. 184.

Chapter 8

The Transfer Matrix

8.1. The Transfer Matrix and the Thermal Zipper

We first encountered the transfer matrix in Sec. 4.9 in connection with the classical configurational partition function for interacting particles in 1D. This methodology really comes into its own in 2D, where it allows a number of classical configurational partition functions to be reëxpressed in terms of the lowest eigenvalue of a solvable *one-dimensional* quantum system.[a] Ising's ferromagnet, studied later in this chapter, is the "cleanest" and most successful example of this technique but we discuss other, simpler, applications as well, starting with the "thermal zipper" of this section.

Consider a two-dimensional square ("*sq*") grid, with objects at sites (n, m), as shown in Fig. 8.1, and interactions along the connecting vertical or horizontal bonds. We wish to sum (or integrate) the Boltzmann factor over the 6 internal degrees of freedom $(\mathbf{p}_{n.m}, \mathbf{r}_{n.m})$ of each object. Because of the interactions, the task is usually quite onerous. For this reason we limit the discussions to *classical* statistical mechanics, in which the momenta and coördinates are variables that can be independently specified. Similarly, in those classical magnetic systems we are able to solve, the spins will be vector quantities of fixed lengths[b] pointing along precisely specifiable directions.

The *zipper* illustrated above is an amusing, nontrivial, system with which to start. Vectors $\mathbf{r}_{n,m} = (x_{n,m}, y_{n,m}, z_{n,m})$ measure the *deviation* of the "particle" on the "*n-m*"th site from its ideal position on the rectangular grid. The "vertical" bonds $K/2 \times (\mathbf{r}_{n,m} - \mathbf{r}_{n,m\pm1})^2$ (i.e. those connecting

---

[a]The transfer matrix can also be used to map problems in 3D statistical mechanics onto 2D quantum systems, but unfortunately the latter are generally as difficult to solve as the former is to evaluate. . .

[b]In the so-called "spherical model", even this mild condition is relaxed.

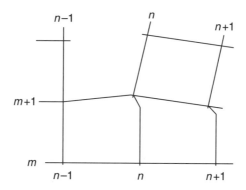

Fig. 8.1. Model for a thermal "Zipper".

Shown: a rectangular grid of N columns $\times M$ rows. The atoms are on vertices connected to their nearest-neighbors (thick lines). Vertical bonds are "unbreakable" harmonic oscillators while horizontal bonds are inelastic and can be torn open. The zipper is shown opening up between nth and $n+1$st column above the $m+1$st row.

nearest-neighbors along the same columns) are elastic and *unbreakable*, as there is a restoring force no matter how great the separation. The "horizontal" bonds denoted U are $U_{n,n-1;m} = -V_0$ if $|r_{n,m} - r_{n-1,m}| < a$ and 0 otherwise; if stretched sufficiently they "break" and the 2D fabric is rent into separate, disconnected vertical strings.

Let us evaluate a typical *vertical* bond $K/2(\mathbf{r}_{n,m} - \mathbf{r}_{n,m-1})^2$ by integrating over the coördinates of the $(n, m-1)$st site. For typographical simplicity write $\mathbf{r}_{n,m} \equiv \mathbf{r}$ for the "target" variable and $\mathbf{r}_{n,m-1} \equiv \mathbf{r}'$ for the dummy of integration. Of course, prior integrations over the rows $m-2$, $m-3$, ... have resulted in a non-negative but otherwise arbitrary function $\Psi(\mathbf{r}_{n,m-1}) \equiv \Psi(\mathbf{r}')$. It, too, appears in the integral shown below. Now, the following *identity*

$$\int d^3r' \left\{ \exp \frac{-K(\mathbf{r} - \mathbf{r}')^2}{2kT} \right\} \Psi(\mathbf{r}') \equiv \left(\frac{2\pi kT}{K} \right)^{3/2} e^{\left(\frac{kT}{2K}\right)\nabla^2} \Psi(\mathbf{r}) \qquad (8.1)$$

proves helpful, as it replaces the Gaussian kernel in this integral expression by a differential operator acting on the target site's coördinate. This identity is verified below.

Problem 8.1.   Let $\Psi(\mathbf{r}') = \exp i\mathbf{q} \cdot \mathbf{r}'$ and show that both sides of the above equation yield identical results. Using Fourier's theorem, show that this identity holds for any arbitrary function Ψ.

Thus, the following *operator* V_1 is the functional equivalent to integrating over *all* N vertical bonds connecting the $m - 1$st row to the mth. It acts on a normalized function Ψ of the positions of particles in the mth row:

$$V_1 = \left(\frac{2\pi kT}{K}\right)^{\frac{3N}{2}} e^{(\frac{kT}{2K})(\sum_{\text{all } n=1}^{N} \nabla_n^2)}. \tag{8.2a}^c$$

We find it necessary also to include a factor V_2 with which to account for the horizontal "zipper" bonds situated *within* the mth row:

$$V_2 = e^{-\beta \sum_n U(r_{n+1} - r_n)}. \tag{8.2b}^c$$

Combining the two we obtain

$$V_1 V_2 \Psi_m(\mathbf{r}_1, \ldots, \mathbf{r}_N) = z\Psi_{m+1}(\mathbf{r}_1, \ldots, \mathbf{r}_N).$$

If the Mth row is connected back to the first the system is translationally invariant in the vertical direction.[d] The largest eigenvalue of a translationally invariant, positive, operator belongs to a translationally invariant, positive, eigenfunction. Thus $\Psi_{m+1} = \Psi_m = \Psi$, is independent of row and,

$$V_1 V_2 \Psi(r_1, \ldots, r_N) = z\Psi(r_1, \ldots, r_N), \quad \text{where } Q = z^M. \tag{8.3}$$

Formally, this same result can be obtained by recognizing that the trace over the variables in the mth row is an inner product connecting matrix operators in the mth and $m + 1$st rows. The configurational partition function Q is:

$$Q = tr\{\cdots (V_1 V_2)_{m-1} \cdot (V_1 V_2)_m \cdot (V_1 V_2)_{m+1} \cdots\}$$

Next, introduce a transformation $\exp S$ with which to diagonalize $V_1 V_2$,

$$Q = tr\{\cdots\} \cdot (e^{-S} V_1 V_2 e^S) \cdot (e^{-S} V_1 V_2 e^S) \cdot (\cdots\}$$

Each diagonal $V_1 V_2$ has a set of eigenvalues $z_0 > z_1 > \cdots$ (listed in descending order), and thus $Q = \sum_{j \geq 0} z_j^M = (z_0)^M [1 + \sum_{j \geq 1} (z_j/z_0)^M] \Rightarrow (z_0)^M$ in $\lim \cdot M \to \infty$. Thus, in the thermodynamic limit, only the largest eigenvalue z_0 contributes to Q. This allows the identification of z in Eq. (8.3) as z_0, the *largest* eigenvalue of $V_1 V_2$, in accord with the general thermodynamic principle that — given a choice — the correct Q is always the largest, ensuring the lowest possible, stablest, thermodynamic free energy.

---

[c]The row index m is common to all the dynamical variables and is omitted henceforth from the operators and coördinates, for typographical simplicity.
[d]The numbers N and M are respectively the total number of sites along each row or column and are ultimately taken to the thermodynamic limit.

Because of the exponentiated sums in V we infer z takes the form $\exp \alpha N$, where α is some constant, hence $\log Q \propto NM$. This shows that the configurational free energy is extensive, as it should be. There is one additional bonus; according to a theorem of Frobenius:

The eigenfunction $\Psi_0(\{\mathbf{r}_n\})$ corresponding to the largest eigenvalue z_0 of a positive operator (or of a matrix with only non-negative elements) has no nodes; hence it is (or can be made) > 0.

It is therefore natural that, after being appropriately normalized, Ψ_0 (*not* $|\Psi_0|^2$) will play the rôle of a *probability* function, the so-called *reduced density matrix* governing the correlations among all $3N$ dynamical variables of a single row, $\mathbf{r}_1, \ldots, \mathbf{r}_N$, that constitute its arguments.

8.2.   Opening and Closing a "Zipper Ladder" or Polymer

In the attempt to gauge the difficulty in solving the zipper problem, we rewrite the exponent in V_1 as:

$$\sum_n \nabla_n^2 \equiv \sum_n \frac{1}{2}(\nabla_n^2 + \nabla_{n+1}^2)$$

and consider the eigenvalue equation governing a single term in this sum, i.e. we study the single bond $\mathbf{r}_{n+1} - \mathbf{r}_n$ (denoted $\mathbf{r}$):

$$\left(\frac{2\pi kT}{K}\right)^{\frac{3}{2}} e^{(\frac{kT}{2K})\nabla^2} e^{-\beta U(r)} \Psi(\mathbf{r}) = \hat{z}^{1/N} \Psi(\mathbf{r}). \tag{8.4}$$

The solution to Eq. (8.4) does not actually solve (8.3), as the exponentiated operators $\frac{1}{2}(\nabla_n^2 + \nabla_{n+1}^2)$ do not commute with functions of $\mathbf{r}_{n+2} - \mathbf{r}_{n+1}$ and $\mathbf{r}_n - \mathbf{r}_{n-1}$. However, one may view (8.4) as the eigenvalue problem of a two-column "ladder" ($N = 2$) with breakable rungs, i.e. of a *two-stranded cross-linked polymer*.

The nodeless solution of (8.4) has to be spherically symmetric, i.e. $\Psi(r) = \varphi(r)/r$. Expressing $\hat{z}^{1/N}$ as $(2\pi kT/K)^{\frac{3}{2}} \exp \beta w$ one is first tempted to combine the exponents and solve a "pseudo-Schrödinger" radial equation for φ,

$$-\frac{\hbar^2}{2m^*} \frac{d^2\varphi(r)}{dr^2} + U(r)\varphi(r) = -w\varphi(r) \tag{8.5}$$

for its *lowest* bound state (i.e. the state with greatest binding energy w). Here $m^* \equiv \frac{\hbar^2 K}{(kT)^2}$. The equivalent problem in elementary quantum theory is solved as follows. The boundary condition $\varphi(0) = 0$ allows only the solution

$\varphi = A \sin kr$ within the spherical well. For $r > a$, outside the well, a second boundary condition $\varphi(\infty) = 0$ admits only $\varphi = Be^{-qr}$ as a permissible solution. A, B are two constants and k is related to q by $\frac{\hbar^2}{2m^*}(q^2 + k^2) = V_0$. The usual conditions that both φ and $d\varphi/dr$ be continuous at $r = a$ yield $ka \cot ka = -qa$, and a condition for the existence of one or more bound states, $V_0 > \frac{\hbar^2\pi^2}{4m^*a^*}$. Recalling the definition of m^* we infer there is no bound state in the high-temperature range, i.e. the binding energy vanishes, $w \equiv 0$, if $T > T_c$, where

$$T_c = \frac{2}{\pi k_B}\sqrt{2Ka^2V_0}. \tag{8.6}$$

Although the one-bond zipper is "broken" above T_c it does heal below T_c. The binding energy, w, the negative of the singular part of the configrational free energy, is plotted in Fig. 8.2. As the temperature is lowered below T_c, w rises quadratically from zero. This signals a second-order phase transition at T_c.

This conclusion is somewhat disturbing, given that the two-stranded molecule is essentially a one-dimensional entity performing a random walk in 3D. It is generally thought that thermodynamical systems do not exhibit a phase transition at any finite T in 1D. But is it wrong?

To check whether the preceding is, in fact, valid we next solve for the *exact* eigenvalues $\hat{z}_j^{1/N} = (\frac{2\pi kT}{K})^{\frac{3}{2}}\exp\beta w_j$ of Eq. (8.4).

Spherically symmetric solutions can be analyzed with the aid of the following lemma:

$$\nabla^2 \frac{\varphi(r)}{r} = \frac{1}{r}\frac{d^2}{dr^2}\varphi(r) \equiv \frac{\varphi^{(2)}(r)}{r};$$

thus for all positive integer n,

$$\nabla^{2n}\frac{\varphi(r)}{r} = \frac{1}{r}\frac{d^{2n}}{dr^{2n}}\varphi(r) \equiv \frac{\varphi^{(2n)}(r)}{r}. \tag{8.7}$$

(The proof is by induction.)

The following variational principle yields w as the extremum of the *rhs*:

$$e^{\beta w} = \frac{\int_0^\infty dr\varphi(r)e^{-\frac{1}{2}\beta U(r)}e^{\beta\frac{\hbar^2}{2m^*}\frac{d^2}{dr^2}}e^{-\frac{1}{2}\beta U(r)}\varphi(r)}{\int_0^\infty dr\varphi^2(r)}. \tag{8.8a}$$

Here we have expressed V_1V_2 in the more symmetric form: $V_2^{1/2}V_1V_2^{1/2}$. The requirement that w be maximal against arbitrary variations $\delta\varphi$ of φ generates the following eigenvalue equation,

$$e^{\beta w}\varphi(r) = e^{-\frac{1}{2}\beta U(r)}e^{\beta\frac{\hbar^2}{2m^*}\frac{d^2}{dr^2}}e^{-\frac{1}{2}\beta U(r)}\varphi(r)\,,\tag{8.8b}$$

identical, to within a similarity transformation, to the original (8.4) (after it is simplified with the aid of (8.7)).

Problem 8.2.   Expanding $\varphi(r)$ in the exact (but unknown) eigenstates Ψ_j of (8.8b) belonging to w_j ($w_0 > w_1 > \cdots$) show that the *rhs* of (8.8a) is an absolute maximum when $\varphi = \Psi_0$. This proves the largest *rhs* ratio yields the desired $w_0 = \text{Max}\{w\}$.

We know the solution of (8.8b) in the two regions: $\varphi(r) = \sin kr$ for $0 < r < a$ and $D \sin ka \exp{-q(r-a)}$ for $r > a$ respectively. Absent any requirement that φ or its derivative be continuous, we must allow D to be arbitrary (albeit, real). The ordinary condition $\frac{\hbar^2}{2m^*}(q^2 + k^2) = V_0$ ensures that $w_0 \equiv \hbar^2 q^2/2m^*$ is the same in the two regions. To determine $\varphi(r)$ near the discontinuity at $r = a \pm \varepsilon$ and to ensure that w is constant there too, we next integrate the two sides of Eq. (8.8b) from $a - \varepsilon$ to $a + \varepsilon$ and proceed to the lim. $\varepsilon \to 0$.

$$\lim \cdot (\varepsilon \to 0) \int_{a-\varepsilon}^{a+\varepsilon} dr\, e^{\beta w}\varphi(r) = O(\varepsilon) \to$$

$$0 = \int_{a-\varepsilon}^{a+\varepsilon} dr\, e^{-\frac{1}{2}\beta U(r)}e^{\beta\frac{\hbar^2}{2m^*}\frac{d^2}{dr^2}}e^{-\frac{1}{2}\beta U(r)}\varphi(r)$$

$$= D(\sin ka)e^{qa}\sum_n \frac{(\beta\frac{\hbar^2}{2m^*})^n}{n!}\frac{d^{2n-1}}{dr^{2n-1}}e^{-qr}$$

$$- e^{\beta V_0}\sum_n \frac{(\beta\frac{\hbar^2}{2m^*})^n}{n!}\frac{d^{2n-1}}{dr^{2n-1}}\sin kr\big|_{r=a}$$

Upon summing the individual series we obtain the expression

$$D = -\frac{qa}{ka}\cot ka\,\frac{e^{q\beta V_0} - e^{\beta\hbar^2(qa)^2/2m^*a^2}}{e^{\beta\hbar^2(qa)^2/2m^*a^2} - 1}\tag{8.9a}$$

that allows the eigenvalue equation for φ to be satisfied everywhere, including at the discontinuity at $r = a$. Now let us insert this eigenfunction into the variational principle, Eq. (8.8a).

We wish the numerator to be proportional to the denominator, with a constant of proportionality exp βw_0. Once again the infinitesimal neighborhood about a yields an extra contribution, denoted X. The condition for X to vanish is precisely,

$$D^2 = -\frac{qa}{ka} \cot ka \frac{e^{q\beta V_0} - e^{\beta \hbar^2 (qa)^2/2m^* a^2}}{e^{\beta \hbar^2 (qa)^2/2m^* a^2} - 1}. \qquad (8.9b)$$

In order for the two expressions (8.9a and 8.9b) to be compatible, D *must* be set = either 0 or 1. The first choice is unphysical, while the second *causes* $\varphi(r)$ *to be continuous* (as we might have guessed all along!) However, unlike in quantum theory, here there is *no* requirement for $d\varphi/dr$ to be continuous at $r = a$. Instead, after setting $D = 1$ we are required to satisfy (8.9a or 8.9b).

In the required *nodeless* solution, ka is limited to the range $\frac{\pi}{2} \le ka \le \frac{\pi}{2}$. Then Eqs. (8.9) (with $D = 1$), augmented by the constant energy condition

$$\frac{\hbar^2}{2m^*}(q^2 + k^2) = V_0,$$

are solved simultaneously. Here the calculations are much less transparent than before and the precise expressions for w and for the related configurational free energy do differ somewhat from what is graphed in Fig. 8.2. Still, the phase transition is confirmed to be of second order and to occur *at the identical* T_c given in Eq. (8.6).

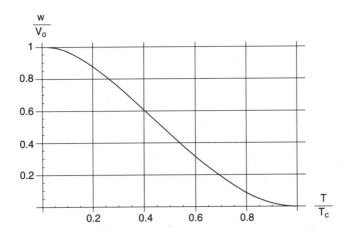

Fig. 8.2. "Binding energy" w/V_0 (i.e. $(-1) \times$ configurational free energy) at temperatures $T < T_c$.

Solution of Eq. (8.5) vanishes as $(T_c - T)^2$ near T_c and is $\equiv 0$ at $T > T_c$.

In only one important detail do the exact solutions differ from those of Eq. (8.5). When solved properly, the exact equations reveal 1 additional, nontrivial, quasi-bound state solution at $w = 0$ from T_c down to $1/2\ T_c$. Below $1/2\ T_c$ it disappears.

Although the $w = 0$ solution could never be accessed in thermodynamical equilibrium, its mere existence allows for *hysteresis* when the zipper ladder is rapidly cooled through the critical point starting from some temperature $T > T_c$ and subsequently re-warmed from below $1/2\ T_c$. Such hysteresis is unusual in a second-order transition — but then, so is the model!

8.3.   The Full Zipper ($N > 2$)

Setting $\Psi(r_1,\ldots,r_N) = \prod_n \frac{\psi(|r_n - r_{n-1}|)}{|r_n - r_{n-1}|}$ in the form of a trial function incorporating all nearest-neighbor correlations, we define the variational eigenvalue as

$$z \equiv \left(\frac{2\pi kT}{K}\right)^{\frac{3N}{2}} \times e^{\beta\tilde{w}N} .$$

According to Frobenius' theorem φ should be taken to be nodeless if one is to optimize z. It is then permissible to integrate both sides of the eigenvalue Eq. (8.3) over all the coördinate variables without risking to obtain zero on either or both sides:

$$\prod_n \int d^3r_n e^{(\frac{kT}{2K})\nabla_n^2} e^{-\beta\Delta U(r_{n+1}-r_n)} \frac{\psi(|r_{n+1} - r_n|)}{|r_{n+1} - r_n|}$$

$$= e^{\beta\tilde{w}N} \prod_n \int d^3r_n \frac{\psi(|r_{n+1} - r_n|)}{|r_{n+1} - r_n|} \tag{8.10}$$

The integrals on the *rhs* are easily decoupled, after specifying the origin of the nth integration (over $\mathbf{r}_n$) to be at $\mathbf{r}_{n+1}$, for $n = 1, 2, \ldots$

On the *lhs* we must be careful to maintain the various factors in their natural order, given that the exponentiated derivatives commute neither with the potential energy terms nor with the ψ/r factors in Ψ. We therefore order the operators on the lhs as follows:

$$\cdots \int d^3r_{n+1} e^{(\frac{kT}{2K})\nabla_{n+1}^2} e^{-\beta U(r_{n+2}-r_{n+1})} \frac{\psi(|r_{n+2} - r_{n+1}|)}{|r_{n+2} - r_{n+1}|}$$

$$\times \int d^3r_n e^{(\frac{kT}{2K})\nabla_n^2} e^{-\beta U(r_{n+1}-r_n)} \frac{\psi(|r_{n+1} - r_n|)}{|r_{n+1} - r_n|} \cdots$$

and once again shift the origin of the nth integration for each $\mathbf{r}_n$ to $\mathbf{r}_{n+1}$. This causes the various integrals to decouple, just as they did on the *rhs*. Extraction of the Nth root of Eq. (8.10) yields,

$$\int_0^\infty d^3r\, e^{(\frac{kT}{2K})\nabla^2} e^{-\beta U(r)} \frac{\psi(r)}{r} = e^{\beta\tilde{w}} \int_0^\infty d^3r \frac{\psi(r)}{r} . \tag{8.11}$$

Certainly the binding energy $\tilde{w}$ is optimized by the "best" solution to this equation, *viz.*, $e^{\beta\tilde{w}}$

$$e^{(\frac{kT}{2K})\nabla^2} e^{-\beta U(r)} \frac{\psi(r)}{r} = e^{\beta\tilde{w}} \frac{\psi(r)}{r} . \tag{8.12}$$

If we now identify $\frac{\psi}{r}$ with $e^{\frac{1}{2}\beta U(r)}\varphi(r)$, Eq. (8.12) becomes formally identical with Eqs. (8.4) and (8.8b), the subject of the preceding section. We can only conclude that $\tilde{w} = w$ to a good approximation and that a second-order, possibly hysteretic, phase transition occurs at or near T_c, the value of which is given in Eq. (8.6) to good accuracy.

8.4. The Transfer Matrix and Gaussian Potentials

Here we examine the amusing example of $U(r) = \frac{1}{2}K'r^2$ in the preceding model, i.e. the special case of horizontal links that are unbreakable and harmonic, just like the vertical links. The thermodynamics in this model are already known and trivial (cf. the Dulong–Petit law) and the spectrum of normal modes is easily obtained. Still, the evaluation of the transfer matrix poses a special challenge. One needs to combine exponentials such as A, B,

$$\mathbf{\Omega}_x \equiv e^{A\frac{\partial^2}{\partial x^2}} e^{-Bx^2} . \tag{8.13a}[e]$$

We give a passing nod to this mathematical challenge (academic though it may be) as ultimately it may spawn practical applications.

First, it is necessary to symmetrize Ω:

$$\Omega_{\text{sym}} \equiv e^{-\frac{1}{2}Bx^2} e^{A\frac{\partial^2}{\partial x^2}} e^{-\frac{1}{2}Bx^2} . \tag{8.13b}$$

After a number of algebraic manipulations one finds,

$$\Omega_{\text{sym}} \Rightarrow e^{-w(a^+ a + \frac{1}{2})}$$

where $w = 2\log(\frac{1}{\lambda_0})$, $\lambda_0 = \sqrt{1+AB} - \sqrt{AB}$, and the eigenvalues of $a^+ a$ are $n = 0, 1, 2, \ldots$. Because Ω_{sym} generates a geometric series

---

[e]Assuming the originating operators to be separable in x, y, and z.

it is easily resummed:

$$\text{Tr}\{\Omega\} = \text{Tr}\{\Omega_{\text{sym}}\} = \frac{1}{2\sqrt{AB}} \, .$$

Exercise for the reader. To arrive at these results one must prove a Lemma:

$$e^{A\frac{d^2}{dx^2}} e^{-Fx^2} = \frac{1}{\sqrt{1+4FA}} e^{-\left(\frac{Fx^2}{1+4FA}\right)} \, .$$

Prove, or verify, this Lemma.

8.5.   Transfer Matrix in the Ising Model

In the Ising model we start by examining functions G of the spin variable S_n, given that $S_n = \pm 1$. An arbitrary function G can *always* be expanded as $G(S_n) \equiv a + S_n b$, with a the even part of G in S_n and b the odd. E.g.:
$\log(1 + \alpha S_m) = \log\sqrt{1-\alpha^2} + S_m \log\sqrt{\frac{1+\alpha}{1-\alpha}}$.

The Hamiltonian for the nearest neighbor two-dimensional Ising ferromagnet is,

$$H = -J\sum_{(i,j)} S_i S_j \, , \tag{8.14}$$

where (i,j) refers to nearest neighbors on an N by M lattice as before. The partition function Z consists of the sum of $\exp -\beta H$ over all 2^{NM} spin configurations. In the spirit of the present chapter we evaluate Z using the transfer matrix.

The key to the thermodynamic properties of the two-dimensional zipper was the integral identity Eq. (8.1). Here, with $K \equiv J\beta$ assumed > 0, it is the *trace* identity,

$$tr_{S'}\{\exp(KSS')G_b(S')\} = 2(a\cosh K + bS\sinh K) \tag{8.15}$$

where by $tr_{S'}$ we mean the sum over the two values ± 1 of S' and continue to use $G_b(S_n) \equiv a + S_n b$. This identity can also be written, $tr_{S'}\{\exp(KSS')G_b(S')\} = (2\cosh K)G_{b'}(S)$, where $b' = b\tanh K$.

Problem 8.3.   Using $tr_{S'}\{1\} = 2$ and $tr_{S'}\{S'\} = 0$, prove (8.15).

In the spirit of Eq. (8.1) we next construct a *local* operator (one for each vertical strand) that will reproduce (8.15). Denote this operator $\Gamma(G)$. It must yield the *rhs* of (8.15) for arbitrary $G(a, b)$:

$$\Gamma(G(S)) = \Gamma(a + bS) \equiv 2(a \cosh K + bS \sinh K). \qquad (8.16)$$

Because G can only assume one of the 2 possibilities, $G = a \pm b$, depending on whether $S = 1$ or -1, it can alternatively be written as a 2-component "*spinor*": $G = [a + b, a - b]$. In this language Γ is a 2×2 matrix and (8.16) is equivalent to the following 2 equations, in matrix notation and using "." to indicate ordinary matrix multiplication:

$$\begin{bmatrix} \Gamma_{11} & \Gamma_{12} \\ \Gamma_{21} & \Gamma_{22} \end{bmatrix} \cdot \begin{bmatrix} a + b \\ a - b \end{bmatrix} = \begin{bmatrix} 2a \cosh K + 2b \sinh K \\ 2a \cosh K - 2b \sinh K \end{bmatrix}. \qquad (8.17)$$

We solve for the matrix elements Γ_{ij}:

$$\Gamma = \begin{bmatrix} \Gamma_{11} & \Gamma_{12} \\ \Gamma_{21} & \Gamma_{22} \end{bmatrix} = \begin{bmatrix} e^K & e^{-K} \\ e^{-K} & e^K \end{bmatrix} = 1 e^K + \sigma_x e^{-K}, \qquad (8.18a)$$

where 1 is the unit 2×2 matrix and σ_x is the first of the three Pauli matrices (each transforming as the component of a vector in 3D).

$$\sigma_x = \begin{bmatrix} 0 & 1 \\ 1 & 0 \end{bmatrix}, \quad \sigma_y = \begin{bmatrix} 0 & i \\ i & 0 \end{bmatrix}, \quad \sigma_z = \begin{bmatrix} 1 & 0 \\ 0 & 1 \end{bmatrix},$$

These matrices are unitary, Hermitean, and *idempotent* ($\sigma_j^2 = 1$) and, together with the unit matrix, form a complete set in which to expand *any* 2×2 operator. In particular, for ferromagnetic coupling $K > 0$ we can *also* write the *logarithm* of Γ in terms of these operators:

$$\Gamma = A e^{\tilde{K} \sigma_x}, \text{ where } A = (2 \sinh 2K)^{\frac{1}{2}} \text{ and } \tanh \tilde{K} = e^{-2K}. \qquad (8.18b)$$

Problem 8.4. (a) Show that $e^{B\sigma_j} = \cosh B + \sigma_j \sinh B$ for $j = x, y$, or z. Using this, derive the expressions for Γ in terms of A and $\tilde{K}$ as given above, assuming K is positive. Then, given (8.18b), prove the so-called "duality relations" (b) $\tanh K = e^{-2\tilde{K}}$ and (c) $\sinh 2\tilde{K} \sinh 2K = 1$.

Problem 8.5. Derive the applicable exponentiated expressions replacing (8.18b) for Γ for *anti*ferromagnetic coupling $\beta J \equiv K < 0$.

In the following we continue to assume $\beta J \equiv K > 0$, i.e. we consider only ferromagnetic, nearest-neighbor, bonds. The figure below illustrates the dual K's of Eq. (8.18b).

For N vertical strands, the "local" part of the transfer matrix is the product of (8.18) over all vertical chains:

$$V_1 = \prod_j \Gamma_j = \prod_j (Ae^{\tilde{K}\sigma_{x,j}}) \tag{8.19}$$

where $\sigma_{x,j}$ refers to the σ_x matrix on the jth vertical strand on the mth row; it commutes with all σ operators on the other strands. Because all operators now refer to the mth row only we can omit the index m. As with the zipper, we must also include into the transfer operator the horizontal bonds within the same row:

$$V_2 = e^{K \sum_j \sigma_{z,j+1}\sigma_{z,j}} . \tag{8.20}$$

This way of writing V_2 may require elaboration. We did *not* write $V_2 = e^{K \sum_j S_{j+1} S_j}$ for we don't immediately know how these spins related to the σ'_xs in V_1. Consider just the jth bond in V_2: $e^{KSS'} = \cosh K + SS' \sinh K$. This 4-valued expression (S and S' can each assume a value ± 1) can be considered to be the diagonal element of a *matrix* operator:

$$\langle mm'|e^{K\sigma_z\sigma'_z}|m'm\rangle = \cosh K + \langle mm'|\sigma_z\sigma'_z|m'm\rangle \sinh K \tag{8.21}$$

in which the 4 possible states $|m'm\rangle$ are m' and m both "up", i.e. $|\uparrow\uparrow\rangle$, and the three other configurations: $|\uparrow\downarrow\rangle, |\downarrow\uparrow\rangle$ and $|\uparrow\downarrow\rangle$. This operator representation turns out extraordinarily fruitful.

Once V_1 and V_2 are formulated in terms of Pauli operators, it becomes apparent that they do not commute.[f] The product of the two non-commuting operators V_1V_2 (or, better, their symmetric product, e.g. $V \equiv V_1^{\frac{1}{2}}V_2V_1^{\frac{1}{2}}$) constitutes the *transfer matrix of the ferromagnetic Ising model* on the *sq* lattice. It has dimension 2^N, large enough to discourage brute-force numerical solutions of the relevant secular determinant. But the *form* of V is simple enough to yield an exact solution — even in the large N limit. However, before going down that road let us start modestly with $N = 2$.

---

[f]But then, neither did V_1 and V_2 in Eq. (8.2) of the zipper problem.

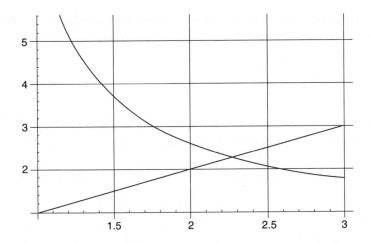

Fig. 8.3. Plot of $1/\tilde{K}$ and $1/K$ versus $1/K(= kT/J)$.

Intersection gives kT_c/J in the self-dual 2D Ising Model on *sq.* Lattice

8.6. The Ising Ladder[g] or Polymer

The symmetric transfer matrix for *just two* vertical strings (together with the cross-links) is,

$$V = A^2 e^{\frac{\tilde{K}}{2}(\sigma_{x,1}+\sigma_{x,2})} e^{K\sigma_{z,2}\sigma_{z,1}} e^{\frac{\tilde{K}}{2}(\sigma_{x,1}+\sigma_{x,2})} \tag{8.22a}$$

Figure 8.3 indicates that at small K the operator V_1 dominates and that at large K it is V_2. Regardless, at *any* K there is *something* to be gained by performing a rotation R of 90° about the y-axis:

$$V \to RVR^{-1} = A^2 e^{\frac{\tilde{K}}{2}(\sigma_{z,1}+\sigma_{z,2})} e^{K\sigma_{x,2}\sigma_{x,1}} e^{\frac{\tilde{K}}{2}(\sigma_{z,1}+\sigma_{z,2})} . \tag{8.22b}$$

A rotation cannot affect the spectrum of eigenvalues z. In the new representation the 4×4 matrix simplifies: it decouples into two noninteracting 2×2 sub-blocks, one subspace for parallel spins and the other for antiparallel spins. This being a model of ferromagnetism, the ground state eigenvector must have the form:

$$\Psi = c|\uparrow\uparrow\rangle + d|\downarrow\downarrow\rangle, \quad \text{with } c+d=1 \text{ and } c,d \text{ both} > 0. \tag{8.23}$$

[g]Recent soluble problems based on the mathematics of the Ising chain or ladder, in an external transverse magnetic field, include the modeling of *biological evolution*, E. Baake, M. Baake and H. Wagner, *Phys. Rev. Lett.* **78**, 559 (1997) and of *nonequilibrium energy transport*, T. Antal, Z. Racz and L. Sasvari, *ibid.*, 167.

The arrows refer to "up" or "down" orientations relative to the rotated z-axis. After factoring out A^2, the secular determinant is,

$$\det \begin{bmatrix} e^{2\tilde{K}} \cosh K - \lambda & \sinh K \\ \sinh K & e^{-2\tilde{K}} \cosh K - \lambda \end{bmatrix} = 0.$$

The formulas in Problem 8.4 are used next to eliminate $\tilde{K}$. One finds for the larger solution (the one compatible with c, d in Eq. (8.23) both > 0):

$$\lambda = \sqrt{\cosh^2 K + (2 \sinh K)^{-2}} + \sqrt{\sinh^2 K + (2 \sinh K)^{-2}} \qquad (8.24)$$

The eigenvalue $z = \lambda A^2$. By inspection, z is analytic in T at all real $T > 0$. We also note that the ladder is ordered at $T = 0$ (all spins $\uparrow$ or all spins $\downarrow$) and disordered at $T \to \infty$. Analyticity precludes a sharp phase transition at any finite T, thus it follows that the order-disorder phase transition is pinned to $T = 0^+$ and that this model ladder is disordered at all finite T.

8.7.   Ising Model on the Isotropic Square Lattice (2D)[h]

We next generalize the transfer operator (8.22b), in the rotated coördinate system, to N columns and M rows:

$$V = A^N e^{\frac{\tilde{K}}{2} \sum_{j=1}^{N} \sigma_{z,j}} e^{K \sum_{j=1}^{N} \sigma_{x,j+1}\sigma_{x,j}} e^{\frac{\tilde{K}}{2} \sum_{j=1}^{N} \sigma_{z,j}}, \qquad (8.25)$$

again omitting, for typographical simplicity, the row index m common to all the spin operators. Let us now reëxpress them in terms of the spin *raising* and *lowering* operators $\sigma^{\pm}$:

$$\sigma_j^+ \equiv \begin{pmatrix} 0 & 1 \\ 0 & 0 \end{pmatrix} \otimes 1 \quad \text{and} \quad \sigma_j^- \equiv \begin{pmatrix} 0 & 0 \\ 1 & 0 \end{pmatrix} \otimes 1 \qquad (8.26)$$

where 1 refers to the unit operator in the space of all spins other than j. The nomenclature derives from the following properties of σ^+:

$$\sigma_j^+ \begin{pmatrix} 1 \\ 0 \end{pmatrix} = \begin{pmatrix} 0 & 1 \\ 0 & 0 \end{pmatrix} \begin{pmatrix} 1 \\ 0 \end{pmatrix} = 0 \text{ and } \sigma_j^+ \begin{pmatrix} 0 \\ 1 \end{pmatrix} = \begin{pmatrix} 0 & 1 \\ 0 & 0 \end{pmatrix} \begin{pmatrix} 0 \\ 1 \end{pmatrix} = \begin{pmatrix} 1 \\ 0 \end{pmatrix};$$

[h]The reader may be interested in the rather convoluted history of Ising's model, skillfully recounted by S. G. Brush in *Rev. Mod. Phys.* **39**, 883 (1967). Primary references for the phase transition include L. Onsager, *Phys. Rev.* **65**, 117 (1944) and Onsager's student Bruria Kaufman, *Phys. Rev.* **76**, 1232 (1949). For the spontaneous magnetization see L. Onsager, *Nuovo Cimento (Suppl.)* **6**, 261 (1949) and C. N. Yang, *Phys. Rev.* **85**, 808 (1952). The present treatment follows the exegesis by T. Schultz, D. Mattis and E. Lieb in *Rev. Mod. Phys.* **36**, 856 (1964), based on the techniques of "fermionization".

and similarly for the lowering operators:

$$\sigma_j^- \begin{pmatrix} 1 \\ 0 \end{pmatrix} = \begin{pmatrix} 0 & 0 \\ 1 & 0 \end{pmatrix} \begin{pmatrix} 1 \\ 0 \end{pmatrix} = \begin{pmatrix} 0 \\ 1 \end{pmatrix} \quad \text{and} \quad \sigma_j^- \begin{pmatrix} 0 \\ 1 \end{pmatrix} = \begin{pmatrix} 0 & 0 \\ 1 & 0 \end{pmatrix} \begin{pmatrix} 0 \\ 1 \end{pmatrix} = 0\,,$$

the spinors referring to the two states of the jth spin. The reader will easily verify the following identities:

$$\sigma_{x,j} = \sigma_j^+ + \sigma_j^- \quad \text{and} \quad \sigma_{z,j} = 2\sigma_j^+ \sigma_j^- - 1\,. \tag{8.27}$$

Rewriting the exponents in (8.25) using (8.27) we recognize them as quadratic forms that normally could be easily diagonalized. (If a and b are either fermions or bosons, expressions such as $\sum_{i,j} a_i Q_{ij} b_j$ are evaluated by diagonalizing the underlying matrix $\mathbf{Q}$.) However, this simplification fails to "work" here.

Spins present a real conundrum. For example, $\sigma_j^+ \sigma_j^- + \sigma_j^- \sigma_j^+ \equiv \{\sigma_j^+, \sigma_j^-\} = 1$, i.e. the Pauli raising and lowering operators *anti*commute at the same site just like the creation and annihilation operators of *fermions*. But at distinct sites it is the *commutator* $\sigma_j^+ \sigma_n^- - \sigma_n^- \sigma_j^+ \equiv [\sigma_j^+, \sigma_n^-] = 0$ that vanishes, as for bosons. Thus, for spins an arbitrary quadratic form cannot be made diagonal by any known, trivial, transformation.

This difficulty could be circumvented if anticommutation relations were imposed for $j \neq n$ as well, i.e. if the spins are *fermionized*. This nonlinear transformation, due to Jordan and Wigner in the 1920's, is not always fruitful. However, in the present instance (where there are only nearest-neighbor bonds), a mapping of spins $\sigma_\pm$ onto fermions c^+ and c *does* pan out. Define:

$$c_j^+ \equiv \sigma_j^+ e^{\pi i \sum_{n<j} \sigma_n^+ \sigma_n^-}\,, \quad c_j \equiv e^{-\pi i \sum_{n<j} \sigma_n^+ \sigma_n^-} \sigma_j^-\,. \tag{8.28}$$

Problem 8.6.   Using the algebra of the Pauli spin operators, prove the following anticommutators for the c_j operators defined in (8.28)

(a) $\begin{cases} \{c_i, c_j\} \equiv c_i c_j + c_j c_i = 0 \\ \{c_i^+, c_j^+\} = 0 \end{cases}$

and   (b) $\{c_i^+, c_j\} = \delta_{i,j}$ $(= 0$ if $i \neq j$ and 1 if $i = j)$.

(c)   Additionally, prove $c_j^+ c_j = \sigma_j^+ \sigma_j^-$, and $e^{2\pi i c_j^+ c_j} = 1$.

Making use of part (c) of Problem 8.6, one inverts (8.28):

$$\sigma_j^+ \equiv e^{-\pi i \sum_{n<j} c_n^+ c_n} c_j^+ \quad \text{and} \quad \sigma_j^- \equiv c_j e^{\pi i \sum_{n<j} c_n^+ c_n}\,. \tag{8.29}$$

Plug the results into the two quadratic forms in the exponentials of V, Eq. (8.25):

$$\sum_j \sigma_{z,j} = \sum_j (2c_j^+ c_j - 1),\qquad (8.30a)$$

and

$$\sum_j \sigma_{x,j+1}\sigma_{x,j} = \sum_j (c_{j+1}^+ + c_{j+1})e^{i\pi c_j^+ c_j}(c_j^+ + c_j)$$

$$= \sum_j (c_{j+1}^+ + c_{j+1})(c_j - c_j^+) \qquad (8.31a)$$

This last is most surprising, but it is the crux of the simplification. Because j, $j+1$ are nearest-neighbors in the numbering scheme all operator phase factors cancel between them except for those relating to j. In the last line of (8.31a) one sees that even these are trivial phases.[i]

Problem 8.7.   Prove $e^{i\pi c^+ c}c^+ = -c^+$ and $e^{i\pi c^+ c}c = +c$.

To simplify the solution further we shall assume periodic boundary conditions (PBC) for the fermions, so as to expand the c, c^+ operators in plane waves:

$$c_n = \frac{1}{\sqrt{N}} \sum_{k\subset BZ} a_k e^{ikn} \quad \text{and similarly for } c^+. \qquad (8.32)$$

The PBC require $\exp ikN = 1$, i.e. $k = 2\pi n/N$ with n an integer in the range $-N/2 \le n < N/2$ (the "first Brillouin Zone" or BZ). The results of Problem 8.6 allow us to recognize the a_k's, as a set of fermions. That is,

$$\{a_k, a_{k'}\} = \{a_k^+, a_{k'}^+\} = 0 \quad \text{and} \quad \{a_k, a_{k'}^+\} = \delta_{k,k'}. \qquad (8.33)$$

Because of translational invariance, the various normal modes decouple — as they did in the weakly interacting Bose gas and in the BCS theory of superconductivity — and the solution of the two-dimensional Ising model becomes "reduced to quadrature". Consider Eq. (8.30a),

$$\sum_j (2c_j^+ c_j - 1) \Rightarrow \sum_{k\subset BZ} (2a_k^+ a_k - 1) \qquad (8.30b)$$

---

[i]For next-nearest-neighbor couplings the phase factors would not have canceled so neatly and the quadratic form in σ_x's could at best only be reduced to a *quartic* form in the fermions.

and (8.31a),

$$\sum_j (c_{j+1}^+ + c_{j+1})(c_j - c_j^+) \Rightarrow \sum_{k \subset BZ} e^{-ik}(a_k^+ + a_{-k})(a_k - a_{-k}^+) \qquad (8.31b)$$

Because this last mixes k and $-k$, we have to combine $k, -k$ into a single sector, i.e.

$$\sum_{k \subset BZ} (2a_k^+ a_k - 1) = \sum_{k \geq 0}^{\pi} 2(a_k^+ a_k + a_{-k}^+ a_{-k} - 1) \qquad (8.30c)$$

and similarly,

$$\sum_{k \geq 0} [e^{-ik}(a_k^+ + a_{-k})(a_k - a_{-k}^+) + e^{ik}(a_{-k}^+ + a_k)(a_{-k} - a_k^+)]$$

$$= \sum_{k \geq 0} 2[\sin k (ia_k^+ a_{-k}^+ + H.C.) + \cos k (a_k^+ a_k + a_{-k}^+ a_{-k} - 1)] \qquad (8.31c)$$

There are 4 possible configurations of fermions at each $k, -k$: the vacuum $|0\rangle$, the one-particle states $a_k^+|0\rangle$ and $a_{-k}^+|0\rangle$, and a doubly-occupied configuration $a_k^+ a_{-k}^+|0\rangle$. The anticommutation relations forbid all other occupancies (algebraically: $(a_k^+)^2 = (a_{-k}^+)^2 = 0$). An obvious symmetry emerges as each quadratic operator separately preserves even or odd occupations. Thus although $|0\rangle$ and $a_k^+ a_{-k}^+|0\rangle$ are connected (as are $a_k^+|0\rangle$ and $a_{-k}^+|0\rangle$), the two sets are disjoint. Therefore, instead of diagonalizing 4×4 matrices we need only diagonalize 2×2 matrices in each even/odd set.

In more detail: Eq. (8.25) takes the form $V = \prod_{k \geq 0} V_k$, where $V_k = A^2 e^{\tilde{K} S_k} e^{2KR_k} e^{\tilde{K} S_k}$, with exponents $S_k = (a_k^+ a_k + a_{-k}^+ a_{-k} - 1)$ and $R_k = [(\sin k)(ia_k^+ a_{-k}^+ + H.C.) + (\cos k)(a_k^+ a_k + a_{-k}^+ a_{-k} - 1)]$.

1. Odd subspaces. If all of the occupation numbers of the k's are in their respective odd-numbered subspaces each of the exponents S_k and R_k is identically zero (by inspection!) The product yields $z = A^N$.
2. Even subspaces. In the subspace $|0\rangle$ and $a_k^+ a_{-k}^+|0\rangle$, we identify $S_k = \sigma_z$ and $R_k = \sigma_y \sin k + \sigma_z \cos k$. The σ's are the Pauli matrices discussed earlier. Here, in the even subspace, V_k has the representation:

$$V_k = A^2 e^{\tilde{K} \sigma_z} e^{2K(\sigma_z \cos k - \sigma_y \sin k)} e^{\tilde{K} \sigma_z}.$$

Being a 2×2 matrix it can be expanded in Pauli operators:

$$V_k = A^2 \begin{pmatrix} e^{2\tilde{K}}(\cosh 2K + \sinh 2K \cos k) & i \sinh 2K \sin k \\ -i \sinh 2K \sinh k & e^{-2\tilde{K}}(\cosh 2K - \sinh 2K \cos k) \end{pmatrix}$$

$$= A^2 \left\{ \begin{array}{l} 1(\cosh 2\tilde{K} \cosh 2K + \sinh 2\tilde{K} \sinh 2K \cos k) \\ \quad + \sigma_z(\cosh 2\tilde{K} \sinh 2K \cos k + \sinh 2\tilde{K} \cosh 2K) \\ \quad - \sigma_y(\sinh 2K \sin k) \end{array} \right\} \qquad (8.34)$$

To simplify further we define an "excitation energy" ε_k,

$$\cosh \varepsilon_k(T) \equiv \cosh 2\tilde{K} \cosh 2K + \sinh 2\tilde{K} \sinh 2K \cos k$$

$$= \cosh 2\tilde{K} \cosh 2K + \cos k \qquad (8.35)$$

(making use of Problem 8.4, part (c).) Next, rotate the z-y plane about the x-axis so as to align the spin operators along the z-axis.

$$V_k = A^2 \{ (\cosh 2\tilde{K} \cosh 2K + \cos k)$$

$$+ \sigma_z \times \sqrt{(\sinh 2\tilde{K} \cosh 2K + \cosh 2\tilde{K} \sinh 2K \cos k)^2 + (\sinh 2K \sin k)^2} \}$$

$$= A^2 \{ \cosh \varepsilon_k(T) + \sigma_z \sinh \varepsilon_k(T) \} . \qquad (8.36a)$$

This expression renders V_k diagonal and reproduces the 2 even-occupancy eigenvalues explicitly. Once V_k is re-exponentiated and σ_z is reassigned its primitive definition S_k (item #2 on preceding page) in terms of fermions, all 4 eigenvalues (not just the even eigenvalues) appear correctly as listed here:

$$V_k = A^2 e^{\varepsilon_k(T)(n_k + n_{-k} - 1)} \quad \text{and} \quad z_k = A^2 \{ e^{\varepsilon_k(T)}, 1, 1, e^{-\varepsilon_k(T)} \} \qquad (8.36b)$$

Here each $n_k \equiv a_k^+ a_k$ is a fermionic "occupation number operator" having two eigenvalue, 0,1. The largest z_k is the first one listed above. The total transfer matrix V and its largest eigenvalue, z, are,

$$V = A^N \prod_{\text{all } k} e^{\varepsilon_k(T)(\frac{1}{2} - \tilde{n}_k)} \quad \text{and} \quad z = e^{N \log A + \frac{1}{2} \sum_{\text{all } k} \varepsilon_k(T)} \qquad (8.37)$$

on the mth row; here $\tilde{n}_k = a_k a_k^+ = 1 - n_k = 0$ in the optimal state. As there are M rows, $Z = z^M$. The free energy is then,

$$F/NM = -kT \left\{ \frac{1}{2} \log(2 \sinh 2K) + \frac{1}{4\pi} \int_{-\pi}^{+\pi} dk \, \varepsilon_k(T) \right\} ,$$

good enough to use, but able to be further simplified:

$$F/NM = -kT\left\{ \log 2 + \frac{2}{(2\pi)^2} \int_0^\pi dk \int_0^\pi dk' \right.$$

$$\left. \log[\cosh^2 2K - \sinh 2K(\cos k + \cos k')] \right\} \qquad (8.38)$$

The last result is obtained using an integral identity given by Onsager,

$$x \equiv \frac{1}{2\pi} \int_0^{2\pi} dk \log[2\cosh x - 2\cos k], \qquad (8.39)$$

and has the advantage of rendering explicit the two-dimensional symmetry inherent in the model.[j] We discuss the consequences of this formula next.

Problem 8.8.   Derive (8.38) using (8.39).

8.8.   The Phase Transition

From inception, $\varepsilon_k(T)$ is symmetric in K and $\tilde{K}$ (cf. definition on p. 238). Interchange of the two equates to a mapping of high- onto low-temperatures, so-called "duality". That is, if we know F at high temperatures $T > T_c$ we can obtain it for low temperatures $T < T_c$ using duality.

If there is a phase transition (and that fact is not yet established!) it can only be located at the intersection of the two regimes at $K = \tilde{K}$, as shown graphically in Fig. 8.3 or found from the solution of $\sinh 2K_c = 1$ (cf. Problem 8.4). The resulting value of T_c is a little more than half the mean-field estimate of $kT_c/J = 4$,

$$kT_c/J = 2/\log(1 + \sqrt{2}) = 2.26919\ldots$$

Let us examine the dispersion $\varepsilon_k(T)$ at T_c and at other temperatures T, whether greater or less than T_c. The following figure illustrates a singular behavior at T_c for long wavelengths.

At — and *only* at — T_c, ε_k vanishes in lim $\cdot k \to 0$. (It does so linearly with $|k|$.) According to Eq. (8.36b) this causes a degeneracy in the eigenvalue spectrum of $V_{k=0}$ in this limit, which is *directly responsible* for the onset of LRO. Moreover, this dispersion anomaly results in a singularity in

---

[j]Onsager's solution was for the more general case in which horizontal bonds J_x differ from vertical bonds J_y. In this way he could analyze the disappearance of the phase transition when $J_x \to 0$. For explanations cf. Schultz, Mattis and Lieb.[h]

the thermodynamic properties at T_c and signals a phase transition at that temperature.

We can get an idea of the nature of this transition by calculating the specific heat in this model. The internal energy *per* site is $u(T) = (1/NM)\partial(\beta F)/\partial\beta$, and after some algebra is found to be:

$$u = -J \coth 2K \left(1 + \frac{2}{\pi}(2\tanh^2 2K - 1)K_1(q) \right) \qquad (8.40)$$

in which $K = J/kT$ as before. $K_1(q) = \int_0^{\pi/2} d\phi(1 - q^2 \sin^2\phi)^{-\frac{1}{2}}$ is the complete elliptic integral of the first kind, with an argument $q \equiv (2\sinh 2K)/\cosh^2 2K$. It exhibits a logarithmic singularity at $q = 1$. Therefore, so too does the derivative of u, the specific heat $c(T)$, plotted in Fig. 8.5.

8.9.   A Question of Long-Range Order

Evidently, the nearest-neighbor ordering, $\langle S_{m,n+1}S_{m,n}\rangle_{TA}$, correlates with $u(T)$, but then so do the more distant, albeit short-range, correlation functions $C(p) \equiv \langle S_{m,n+p}S_{m,n}\rangle_{TA}$ that decay (to a first approximation) as

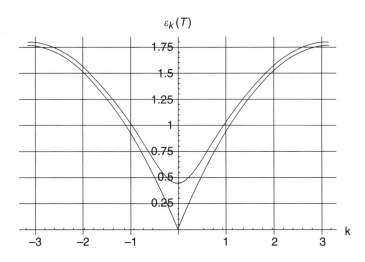

Fig. 8.4.   Dipersion of the eigenvalue $\varepsilon_k(T)$ at T_c (lower curve) and at some $T < T_c$ (or at its dual $T > T_c$) (upper curve).

This figure illustrates two aspects of the exact solution: (1) linear dispersion for small k found at, and *only* at, T_c and (2) a parabolic minimum at $k = 0$ together with a temperature-dependent gap $\varepsilon_0(T)$ at $k = 0$ ("mass-gap") at all other values of T.

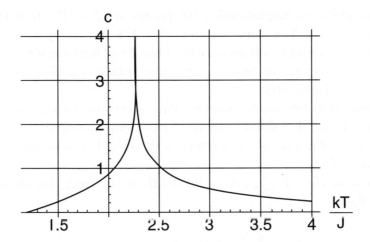

Fig. 8.5. $c(T) = du/dT$ exhibits logarithmic singularity near T_c.
Curve obtained by differentiating (8.40) analytically, using *Mathematica*. (in units $k_B = 1$ and $J = 1$)

$\approx C(1)^{|p|}$ at finite p. For *long-range order (LRO)* no finite p will do. It is necessary to proceed to the limit $p \to \infty$. For consider the (intensive quantity) *magnetization density*, or rather, its square:

$$m^2 = \left(\frac{1}{N}\right)^2 \left\langle \left(\sum_{n=1}^{N} S_{n,m}\right)^2 \right\rangle_{TA}, \tag{8.41}$$

a quantity presumably independent of the row (index m) on which it is evaluated. There are N contributions $C(0) = 1$ that vanish when divided by N^2 in the thermodynamic limit $N \to \infty$ (as do $C(\pm 1)$, $C(\pm 2)$, $C(\pm 3)$, ...etc). If, however, $C(p)$ achieves a non-zero limiting value for $C(\infty)$ at large $|p|$, then $m^2 \to |C(\infty)|^2$ in the thermodynamic limit, a manifestation of *ferromagnetic LRO*.

The calculation of $C(\infty)$ requires the evaluation of determinants of Toeplitz matrices[k] (poor cousins of the determinants of cyclic matrices). After lengthy analysis, an explicit result can be obtained in the present case. It is,

$$m^2 = (1 - 1/\sinh^4 2K)^{\frac{1}{4}} \text{ for } T < T_c \text{ and zero otherwise}, \tag{8.42}$$

[k]See G. Newell and E. Montroll, *Rev. Mod. Phys.* **25**, 353 (1953) or footnote h.

i.e. the spontaneous magnetization disappears as $(T_c - T)^\beta$ near the critical point $T \to T_c$.[1] The same *exponent* $\beta = 1/8$ (not to be confused with $(\beta \equiv 1/kT)$ is found in *all* nearest-neighbor Ising ferromagnets in 2D: on triangular, hexagonal, etc. lattices. The other critical exponents also depend on d but not on geometry.

The related *paramagnetic susceptibility* $\chi_0 = dm/dB|_{B \to 0}$ can also be calculated. Although defined in terms of an external magnetic field $\chi_0(T)$ is actually, like $c(T)$, a physical property of the model even in the absence of an external field.

Given $H = H_I - B \sum_{m,n} S_{m,n}$, where H_I is the Ising Hamiltonian, the susceptibility is:

$$\chi = \frac{1}{NM} \frac{d}{dB} \frac{Tr\{\sum_{m,n} S_{m,n} e^{-\beta H_I + \beta B \sum S_{m,n}}\}}{Tr\{e^{-\beta H_1 + \beta B \sum S_{m,n}}\}}$$

$$= \frac{\left\langle \left(\sum_{m,n} (S_{m,n} - \langle S_{m,n} \rangle_{TA}) \right)^2 \right\rangle_{TA}}{NM \, kT}. \tag{8.43}$$

When evaluated in zero field χ is denoted χ_0 as usual. For the Ising ferromagnet, regardless of whether the lattice is *sq*, hexagonal, triangular, or whatever, one again finds a universal law in 2D:

$$\chi_0 \propto \frac{1}{|T - T_c|^\gamma}, \text{ the critical exponent being } \gamma = 7/4. \tag{8.44}$$

This susceptibility is far more singular near the critical point than is Curie's Law ($\gamma = 1$). Numerical studies of the Ising model in 3D indicate this critical exponent to be somewhat smaller but still $> 1 : \gamma = 5/4$ in 3D, whether the lattice is *sc*, *bcc*, *fcc* or whatever. But as *the number d of dimensions* is increased γ drops to its mean-field value $\gamma = 1$ in all $d \geq 5$.

8.10. Ising Model in 2D and 3D

There exist but a handful of models of interacting particles that can be solved in 3D; unfortunately the Ising model is not among them. A reasonable *guess* might entail generalizing the two-dimensional expression (8.38) to a cube

---

[1]This result was first announced at the blackboard by Onsager, without any explanation; the first published derivation, some years later, is that of C. N. Yang.[h]

$L \times M \times N$ having a cube-like BZ:

$$F/LNM = ? - kT\left\{ \log 2 + \frac{1}{2(2\pi)^3} \int_0^{2\pi} dk \int_0^{2\pi} dk' \int_0^{2\pi} dk'' \right.$$

$$\left. \log[\cosh^2 2K \cosh 2K_z - \sinh 2K(\cos k + \cos k') - \sinh 2K_z \cos k''] \right\}$$

This expression reduces to (8.38) if one sets the new bonds $K_z \to 0$. For the isotropic sc lattice $K_z = K$ this integrand has a singularity at $k = k' = k'' = 0$ at a value of K that satisfies $\cosh^3 2K_c = 3\sinh 2K_c$, i.e. it predicts $kT_c \approx 4.79J$. Expanding the integral about that temperature reveals $\alpha = 1/2$, i.e. a singular specific heat with $c \propto |T - T_c|^{-1/2}$. Is any of this believable? The answer is, *No*.

Accurate series for the various thermodynamic functions — especially the zero-field susceptibility and the specific heat — both diverge explicitly at T_c. These include low-temperature expansions in powers of $\exp -J/kT$ at low T and high-temperature expansions in powers of $\tanh J/kT$ at high temperatures, resummed in irreducible clusters.[m] Comparison of the numerical results with the above guesswork is not favorable to the latter. Unambiguously, the series expansions yield $kT_c \approx 4.5J$ in the sc lattice, lower than the guessed value! In the diamond (*zincblende*) lattice, each spin has 4 nearest-neighbors, just as in the sq lattice solved earlier. But as shown in the figure on the next page, kT_c in this 3D lattice is found to be $\approx 3.4J$, closer to the mean-field value, $4J$, than is $2.3J$ for the 2D lattice with identical coördination number.

As determined from not just series expansions but also from the exact transfer matrix solutions, *all* Ising ferromagnets with short-range unfrustrated interactions on all 2D lattices have a logarithmic singularity in the specific heat. They also share all other *critical exponents* but *not* T_c which varies for each.[n]

In 3D the series expansions on a variety of lattices with n nearest-neighbor Ising interactions reveal an almost mean-field critical temperature $kT_c \approx n \times 3/4J$ ($\pm 20\%$). The thermal exponent α in $c \approx 1/|T - T_c|^\alpha$, has a common value $\alpha = 1/8$ in 3D, far from the initial guess $1/2$. Detailed agreement of the numerical expansions with the experimental data such as that shown in

---

[m]There exists a vast literature and some books on this topic. Some references: M. Sykes *et al.*, *J. Phys.* **A5**, 624*ff*, 640*ff*, 667*ff* (1972), **A6**, 1498, 1506 (1973).

[n]Onsager finds $\arctan[\sinh 2K_c] = p/n$ in the hexagonal, sq, and triangular 2D lattices, i.e. $kT_c = 1.59\ldots$ on the hexagonal lattice ($n = 3$), $2.27\ldots$ on the sq lattice ($n = 4$) and $3.64\ldots$ ($n = 6$) where n = coördination number of nearest-neighbors = 4,6,3 respectively.

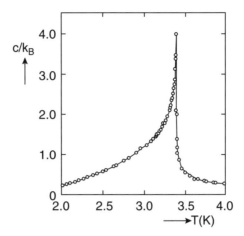

Fig. 8.6. **Specific Heat $c(T)$ of DyPO$_4$ (consisting of Ising Spins on 3D Diamond Lattice): "Exact" Theory (*series expansions*) (—) versus Experiment (ooo).**

The solid curves are well fitted by $\alpha = 1/8$, the thermal exponent expected for 3D Ising lattices.[°]

Fig. 8.6, deals the final *coup de grace* to the guessed integral at the top of the preceding page.

8.11.   Antiferromagnetism and Frustration

As we saw, the *type* of lattice is almost irrelevant in ferromagnetism, but this "detail" becomes important when the interactions favor nearest-neighbors to be antiparallel, as in antiferromagnets. *Bi-partite* lattices (*sq*, hexagonal, *sc* (simple cubic), *bcc* (body-centered cubic)) can be decomposed into two sublattices, such that the spins on the A sublattice interact only with spins on the B sublattice. Then if the sign of J favors antiparallelism, a simple "gauge" transformation on the Ising spins: all $S_i \to -S_i$, performed *only* on the A sublattice, transforms the model into a *ferromagnet in the new* "*pseudospins*". Let us assume this pseudo-ferromagnet to have been solved. Even so, the interaction with an homogeneous external field is changed:

$$+B \left(\sum_{m,n \in A} S_{m,n} - \sum_{m,n \in B} S_{m,n} \right). \qquad (8.45)$$

[°]From L. deJongh and A. Miedema, *Adv. Phys.* **23**, 1–260 (1974).

It favors pseudospins on the two sublattices to be antiparallel, whereas the internal forces now favor parallelism. Because the internal and external order parameters differ, it is possible to impose a finite magnetic field $|B| < B_c(T)$ on an antiferromagnet without destroying the phase transition. The "spin-flop field" B_c is maximum at $T = 0$ and decreases with increasing T, vanishing at T_c.

Problem 8.9. Show: $B_c(0) = |J|/2$ on sq lattice (2D) and sc (3D).

The *triangular* lattice is *not* bipartite: each spin has 6 nearest-neighbors, of which each is itself a nearest-neighbor to 2 out of the 6. Wannier[P] was first to study the ground-state properties of the Ising nearest-neighbor antiferromagnet on this lattice. Instead of a phase transition he found that the high-T phase extends to $T = 0$, where there is a very high ground-state degeneracy. The following are exact results, obtained analytically using the transfer matrix.

The ground-state entropy is $\mathscr{S}_0 = k_B 0.32306\ldots$ *per* spin (approximately half the $T = \infty$ entropy *per* spin, $k_B \log 2 = k_B 0.69315\ldots$; also cf. the entropy at T_c on the sq lattice: $k_B 0.30647\ldots$, or on the hexagonal lattice: $k_B 0.26471\ldots$) A high ground-state entropy is a patent violation of Nernst's theorem; it is the typical consequence of *frustration*. The root cause is this geometry, in which it is impossible to satisfy all AF bonds. But even if the bonds are not all the same, some being antiferromagnetic (J) and others ferromagnetic ($-J$), one can also achieve frustration — as in the following example, illustrated in Fig. 8.7.

An odd number (either 1 or 3) of AF bonds frustrates any given square plaquette. All 3 plaquettes are frustrated in the above example. Moreover in the absence of external fields *they are equivalent*. To verify this, just change the orientation of the z-axis by 180° on sites #1 and #2 only. This gauge transformation cannot affect the spectrum of eigenvalues. All three plaquettes are now formally identical to the bottom one and the ground state energy (given by the configuration of all spins "up") is found by inspection: $E_0 = -4J$. In exchange for a substantial cost in energy, frustration creates a high ground-state degeneracy. This degeneracy is the hallmark of *spin-glasses*, materials in which the nearest-neighbor interactions on a given lattice are random.

[P]G. Wannier, *Phys. Rev.* **79**, 357 (1950); for a review: C. Domb, *Adv. Phys.* **9**, 149–361 (1960). See also Ref. h, p. 196.

Problem 8.10. Find the ground-state degeneracy of the 3-plaquette system illustrated in Fig. 8.7.

8.12. Maximal Frustration

Figure 8.8 is a generalization of Fig. 8.7 to 2D. Too regular to be a spin glass, this is an homogeneous realization of the case of 25% *AF* bonds with no fluctuations in their distribution. Note that adding just 1 more *AF* bond will *decrease* the frustration, as would subtracting just 1. Thus, this is an example of *maximum frustration*. J. Villain called it "a spin-glass without disorder." We can ask some interesting questions of this model: is there an order-disorder phase transition at any finite T? What is the ground state energy? Is the ground state entropy macroscopic? And after solving for the eigenvalues of the transfer matrix, we might wish to obtain the thermal properties of this model and its ground state energy and entropy.

Some of these goals present a challenge, albeit not an insuperable one. Others are easily met and are relegated to the following Problem.

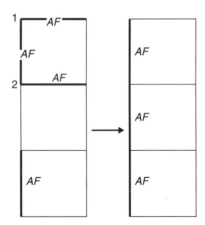

Fig. 8.7.   Illustration of *frustration* on an Ising spin ladder.

Each of the 3 plaquettes shown is frustrated! All are, despite appearances, essentially equivalent. Let the bond between any two neighbors be $\pm JSS'$, with $-J$ for ferromagnetic and $+J$ for *AF*. The Ising spins are ± 1. If the bonds were all $+J$ or all $-J$, the ground state energy of the 9 bonds shown would be $-9J$; as it is, the lowest $E_0 = -4J$.

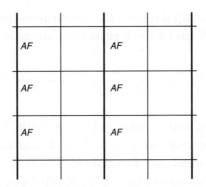

Fig. 8.8.   Maximally frustrated Ising model on the *sq* lattice (2D).
Bonds on every second riser are AF, as indicated; all other bonds are ferromagnetic.
The direction for the transfer matrix is now *horizontal*.

Problem 8.11.   Find the ground state energy of the maximally frustrated
nearest-neighbor Ising model illustrated below. *Estimate*, if you can, the
corresponding ground-state degeneracy.

As will be shown, the model is exactly solvable. We indicate the solution
but because it is so complicated do not carry it out in detail, except for
obtaining a *proof* that there is no phase transition at finite T in this model.

Here the transfer matrix is simpler when it is evaluated along the horizontal. There are 2 columns in each unit cell, so $Z = z^{N/2}$, with z the eigenvalue:
$V_F V_{AF} \Psi = z \Psi$.

As on pp. 199–200, decompose V into k-sectors:

$$V_k = V_{F,k} V_{AF,k} = A^4 e^{2\tilde{K} S_k} e^{2K R_k} e^{2\tilde{K} S_k} e^{-2K R_k} .$$

Inserting parentheses in strategic spots, then performing a similarity
transformation to an explicitly Hermitean form, obtain:

$$V_k = A^4 e^{2\tilde{K} S_k} e^{2K R_k} e^{2\tilde{K} S_k} e^{-2K R_k}$$
$$\Rightarrow A^4 \left(e^{2\tilde{K}(e^{-K R_k} S_k e^{K R_k})}\right) \cdot \left(e^{2\tilde{K}(e^{K R_k} S_k e^{-K R_k})}\right) \qquad (8.46)$$

Rotating c.c.w. by k in the z, y plane transforms S_k and R_k:

$$S_k = \sigma_z \Rightarrow \sigma_z \cos k + \sigma_y \sin k , \qquad R_k = \sigma_z \cos k - \sigma_y \sin k \Rightarrow \sigma_z .$$

The exponents are now:

$$\left(e^{\pm K R_k} S_k e^{\mp K R_k}\right) = \sigma_z \cos k + \sigma_y \sin k \cosh 2K \mp i \sigma_x \sin k \sinh 2K .$$

After some further manipulations, the eigenvalues are found. After being expressed entirely in K and k the largest ones are $z_k = A^4 \exp 2\varepsilon_k(T)$, where

$$\cosh 2\varepsilon_k(T) = 1 + \frac{2}{\sinh^2 2K} \times (1 + \sinh^2 2K \sin^2 k) \qquad (8.47)$$

As T approaches absolute zero, $\varepsilon_k(T) \to (\sinh^{-2} 2K + \sin^2 k)^{1/2}$. The "gap" $1/\sinh 2K$ disappears at $T = 0$. Thus F has an essential singularity (a high-order phase transition), but only at $T = 0$.

Other aspects of this model have been studied extensively.[q]

The concept of *geometrical frustration* extends to phenomena other than antiferromagnetism. These include *ice formation*, surface adhesion, etc., and to exotic materials, e.g. *pyrochlores*. The reader is directed to the pioneering compilation by Ramirez.[r]

Problem 8.12. (a) Derive (8.47). (b) Calculate the ground state energy *per* spin E_0. (c) Obtain the ground state entropy *per* spin $\mathscr{S}_0$ by calculating $(E_0 - F)/NMT$ in the lim $\cdot T \to 0$. above.

8.13.   Separable Model Spin-Glass without Frustration

Because frustration is the joint property of geometry and antiferromagnetism it is expected to play a major rôle in systems with random bonds — especially in *spin-glasses*. In this section we examine simpler counter-examples consisting of systems that are random but devoid of frustration. In some instances their properties can be mapped precisely onto their non-random counterparts. In the following example, the bonds connect only nearest-neighbors and take the *separable* form $J_{ij} = -J\varepsilon_i\varepsilon_j$ where the $\varepsilon_j = \pm 1$ at random ($\langle \varepsilon_i \rangle = 0$). In this model the disorder is *site-centered* rather than *bond-centered* as previously. The Hamiltonian of the nearest-neighbor separable Ising spin-glass in an external field B is:

$$H = \sum_{(i,j)} J_{ij} S_i S_j - B \sum_i S_i \qquad (8.48)$$

[q]cf. G. Forgacs, Phys. Rev. **B22**, 4473 (1980) and references therein.
[r]A. P. Ramirez, *Geometrical Frustration*, Handbook of Magnetic Materials, 1999.

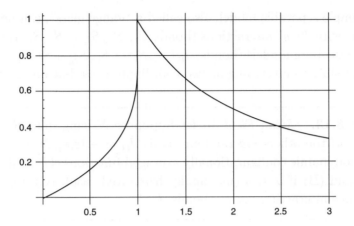

Fig. 8.9.   Zero-field susceptibility $\chi_0(T)$ (arbitrary units) as function of T/T_c.

Calculated for 2D Separable Ising spin-glass, showing the cusp at T/T_c.

After defining the *local* direction at the ith site by ε_i we can replace the spins S_i by a new set $T_i \equiv S_i\varepsilon_i$, also ± 1. Then,

$$H \Rightarrow -J \sum_{(i,j)} T_i T_j - B \sum_i \varepsilon_i T_i. \tag{8.49}$$

In zero external field, the phase transitions is patently identical to that in the corresponding Ising ferromagnet. The zero-field magnetic susceptibility is a different matter. Assuming $\langle \varepsilon_i \rangle = 0$, the configurational averages of $\langle S_i \rangle_{TA} = \varepsilon_i \langle T_i \rangle_{TA} = \varepsilon_i m$ all vanish.

$$\chi_0(T) = \frac{\langle (\sum_i S_i)^2 \rangle_{TA} - \langle \sum_i S_i \rangle_{TA}^2}{kT} = \frac{N}{kT}(1 - \langle m \rangle_{TA}^2) \tag{8.50}$$

where $m = |\langle T_i \rangle_{TA}| \approx (1 - T/T_c)^\beta$ is the spontaneous magnetization below T_c. If χ_0 satisfies Curie's Law $1/T$ for $T > T_c$ and a different law $\approx (T/T_c)^\beta$ below T_c, it must exhibit a discontinuity in slope at T_c proper. Such a cusp, as in Fig. 8.9 is indeed frequently observed in spin glasses.

Combining frustration with randomness lowers T_c or may even eliminate the phase transition altogether. Realistic spin-glasses exhibit hysteresis and multiple other interesting properties[s] while models of spin glasses have found exciting applications.[t]

[s]See K. Binder and A. P. Young, *Rev. Mod. Phys.* **58**, 801–976 (1986).
[t]E.g., code-breaking or error-correcting: H. Nishimori, *Statistical Physics of Spin Glasses and Information Processing*, Oxford Univ. Press, Oxford, 2001.

Our simple separable model, also called a "gauge glass", can be extended
to systems with "x-y" interactions (bonds: $J_{ij}(S_{i,x}S_{j,x}+S_{i,y}S_{j,y})$) but *not* to
"x-y-z" Heisenberg models (bonds: $J_{ij}(S_{i,x}S_{j,x}+S_{i,y}S_{j,y}+S_{i,z}S_{j,z})$), because
it is permissible to invert one or two coördinate axes, but *not all three*.

Problem 8.13.   *Mean-Field Model.* Supppose N Ising spins S_i, each con-
nected to *all* the others *via* random bonds $J_{ij} = -J\varepsilon_i\varepsilon_j/N$, $\varepsilon_i = \pm 1$. Find
the thermodynamic configurationally averaged functions $c(T)$ and $\chi_0(T)$ (A)
if $J > 0$ and (B) if $J < 0$ (maximally frustrated case). (C): repeat (A) &
(B) for spins interacting *via* x-y bonds J_{ij}.[u]

8.14.   Critical Phenomena and Critical Exponents

Over the last few decades a great deal of attention has been paid to *critical
phenomena*, the thermodynamic properties of systems at or near a phase
transition. Such studies were originally informed by "critical opalescence",
in which an ebullient substance refracts all optical wavelengths equally and
there appears to be no characteristic scale or distance. The importance of
the initial studies became magnified by their universal applicability. Whether
a material is a fluid or a ferromagnet, its critical properties can be virtually identical! They depend on such parameters as d (the number of spatial
dimensions), the order parameters (including the number of components,
q, and whether the components transform as components of a vector, or
scalars, etc.), so ultimately, on the *universality class* to which the material
belongs.

　　This field was pioneered in the early 1960's by B. Widom,[v] C. Domb, G.
Rushbrooke and their many students and was brought to general notice in
two articles: one authored by L. P. Kadanoff and students enrolled in his
class at the University of Illinois[w] and the other by M. E. Fisher.[x] While the
theory originated in series expansions it was ultimately reinterpreted on the
basis of *scaling*. Stanley's text[y] surveys the original bases of this topic. Sophisticated derivations based on renormalization group and "ε-expansions"

---

[u]Phase diagram in: P. Sollich *et al*, *J. Phys. Soc. Jpn.* **69**, 3200 (2000).
[v]B. Widom, *J. Chem. Phys.* **43**, 3898 (1965).
[w]L. P. Kadanoff *et al.*, *Rev. Mod. Phys.* **39**, 395–431 (1967).
[x]M. E. Fisher, *Rep. Progr. Phys.* **30**, 615 (1967).
[y]H. E. Stanley, *Introduction to Phase Transitions and Critical Phenomena* (Oxford U.
Press, Oxford 1971).

Table 8.1.   Definitions of Critical Exponents.

Magnetic System	Fluids	Power Law
Zero-field specific heat c_B	Specific Heat at constant volume c_V	$c \propto 1/|t|^\alpha$
Zero-Field Spontaneous Magnetization m	Liquid-gas Difference in Density $\rho_l - \rho_g$	m or $\rho_l - \rho_g \propto (-t)^\beta$
Zero-Field Susceptibility χ_0	Isothermal Compressibility κ_T	χ_0 or $\kappa_T \propto 1/|t|^\gamma$
Correlation Length	Correlation Length	$\xi \propto 1/|t|^\nu$
Spin-Spin Correlation Function at T_c	Density-density Correlation Function at T_c	$G(R) \propto 1/R^{d-2+\eta}$
Critical Isotherm $(t = 0)$	Critical Isotherm $(t = 0)$	$|B| \propto |m|^\delta$ or $|p - p_c| \propto |\rho_l - \rho_g|^\delta$

about 4 dimensions followed.[z] In 2D, all solved models could be related *via* conformal invariance.[aa] A simple up-to-date survey of these topics by Yeomans is available.[bb]

In the following pages we recount only a few results, where they relate to material covered in this and previous chapters. For greater detail we suggest the readings in Refs. v–bb before confronting what has lately become a rather extensive and daunting literature on the subject.

Consider the 2-body correlation function $g(R) = (\exp -R/\xi)/R^{d-2+\eta}$ for $d \geq 2$, in which η is an "anomalous" exponent to fit the actual correlations; the correlation distance $\xi \approx 1/t^\nu (t = T/T_c - 1)$ diverges at the the critical point, i.e. at $T = T_c$ and $B = 0$ in ferromagnetic systems. The *scaling hypothesis* assumes a simple equation of state, $B = m^\delta \Psi(t/m^{1/\beta})$ and a simple form for the singular part of the free energy, $F(t, B) = t^{2-\alpha}\varphi(t/B^{1/\beta\delta})$. All quantities having dimension "length" are proportional to ξ, regardless whether they originate in an external $B \neq 0$ at $t = 0$ or in $t \neq 0$ at $B = 0$.

Consider the 2D Ising model with $g(R_{ij})$ the 2-spin correlation function. The calculated exponent of the magnetization is $\beta = 1/8$, implying $\eta = 1/4$. This and other critical exponents are defined in Table 8.1 and their values

[z]S. J. Amit, *Field Theory, the Renormalization Group and Critical Phenomena* (McGraw-Hill, New York, 1978).
[aa]C. Itzykson, H. Saleur and I.-B. Zuber, *Conformal Invariance and Applications to Statistical Mechanics*, World Scientific, Singapore, 1988.
[bb]J. M. Yeomans, *Statistical Mechanics of Phase Transitions*, Clarendon Press, Oxford, 1992.

Table 8.2. Some Models and their Critical Exponents.

Name or Universality Class	Symmetry	α	β	γ	δ	ν	η	Physical Example
2D Ising	1 component scalar	0	1/8	7/4	15	1	1/4	Hydrogen adsorbed on Fe surface
3D Ising	//	0.1	0.33	5/4	4.8	0.63	0.04	Lattice gas
3D X-Y	2 component vector	0.01	0.34	1.3	4.8	0.66	0.04	Superfluid Superconductor
3D Heisenberg	3 component vector	-0.12	0.36	1.39	4.8	0.71	0.04	Ferro- and antiferro-magnets, ordinary fluids[cc]
Mean-Field	any	0	1/2	1	3	1/2	0	Infinite-ranged ferromagnetic forces; van der Waals fluids

given in Table 8.2. (*N.B.*: when fitted to a power law, both the logarithmic singularity and the discontinuity in the specific heat are accorded the same exponent $\alpha = 0$, this being the closest exponent.)

For the derivations we exhibit some of the so-called *scaling* or *hyperscaling relations* that relate the various exponents in d dimensions:

$$\alpha + 2\beta + \gamma = 2\,, \quad \alpha + \beta(1+\delta) = 2\,, \quad \gamma = \beta(\delta - 1)\,,$$
$$2 - \alpha = d\nu\,, \quad \gamma = (2-\eta)\nu\,. \tag{8.51}$$

Some of these relations were originally expressed as inequalities, the first of which being the subject of the Problem below. The second was originally known as *Griffiths' Inequality*, the fourth one is due to Josephson and the last to Fisher.

For a prototype derivation, recall Eq. (3.16):

$$C_B = C_M + T(\chi_T)^{-1}[\partial m/\partial T|_B]^2$$

[cc]Fluids share with the spin 1/2 ferromagnet a hard-core repulsion, short-range attraction and diffusive motion; therefore logically belong to the same, *universality class*. Recall Guggenheim's plot, Fig. 3.4, indicating $\beta \approx 1/3$ (and *not* 1/2 as predicted by van der Waals) for a variety of fluids.

Set $B = 0$ and omit the specific-heat term on the *rhs*:

$$C_{B=0} \geq T(\chi_T)^{-1}[\partial m/\partial T|_{B=0}]^2 \tag{8.52}$$

Problem 8.14. Prove *Rushbrooke's Inequality*, $\alpha + 2\beta + \gamma \geq 2$, using Eq. (8.52).

Assuming there is a *critical surface* near a critical point, two parameters must govern the approach to the critical point, hence there are just two critical numbers, y_1 and y_2. Whenever this holds, the physical exponents $(\alpha, \beta, \ldots$ *etc.*) except for η (given by the last equation in (8.51),) are all expressible in the two y's, as follows:

$$\alpha = 2 - d/y_1, \quad \beta = (d - y_2)/y_1, \quad \gamma = (2y_2 - d)/y_1, \quad \delta = y_2/(d - y_2)$$

The reader will wish to check the ostensible accuracy of the various Eqs. (8.51), using these data.

8.15. Potts Models

While the 3D Ising model mimics a lattice gas (spin "up" corresponds to an atom, "down" to a vacancy) there is need for a greater number of degrees of freedom to characterize an alloy (e.g.: $q = 3$ to describe atom A, B and vacancy at any given site, or arbitrary q for an alloy composed of q constituents.)

The following model, suggested in the early 1950's by Cyril Domb as a Ph.D. topic for his student R. B. Potts, turned out to be both useful and flexible. The Potts model relates to the Ising model and to alloy formation. In its simplest version it is,

$$H_0 = -2J \sum_{(i,j)} \delta(n_i, n_j) \tag{8.53}$$

Each site is characterized by a coördinate $n_i = 1, 2, \ldots, q$. If the coördinates of two neighboring sites coïncide, $\delta = 1$ and the energy is lowered by $2J$, otherwise $\delta = 0$ and there is no energy gain or loss. Thus the ground state is q-fold degenerate (one of q ferromagnetic-type states in which all sites have the same n_j.) The natural tendency of a system at finite T being to increased disorder, it is clear that there can occur an order-disorder phase transition of *some* sort at *some* temperature T_c which depends on q and d.

Dynamically, the Ising model is the $q = 2$ version of Potts' model, therefore we know that the latter has a second-order phase transition for $q = 2$ in $d \geq 2$ dimensions. But what is the nature of the phase transition, if any, for $q \geq 3$ in $d \geq 2$ dimensions? And what of the many generalizations of the Potts model? For example,

$$H = - \sum_{(i,j)} 2J(n_i)\delta(n_i, n_j) - \sum_i \sum_m b(m)\delta(n_i, m). \qquad (8.54)$$

What of the "antiferromagnetic" versions? There exists a famous coloring problem: how many ways are there to color a map such that no two adjacent countries are the same color? (Answer: it is given by the ground state entropy for $J < 0$.) There is available an excellent review that deals with the many physical aspects of this topic[ee] and an extensive mathematical/analytical book on the subject.[ff]

As in other applications the transfer matrix approach to solving Potts' model is at its most useful in 2D although it provides only a starting point in higher dimensions. Using algebraic duality it is possible to pinpoint the phase transition where there is one. An abbreviated construction of the transfer matrix of the two-dimensional model and derivation of the duality relation follow.

The two conjugate variables are n_j and $p_j = -i\partial/\partial n_j$ each spanning q values at each site: $n = 0, 1, \ldots, q-1$ and the eigenvalues of p being $0, \pm 2\pi/q, \ldots$ The unsymmetrized transfer operator is $V = V_1 V_2$ as usual, with $V_2 = \exp 2\beta J \sum_j \delta(n_j, n_{j+1})$ providing the Boltzmann factors for all the bonds within the mth row. V_1 is once again replaced by a local operator that mimics the trace operation on a vertical bond connecting a site of the $m - 1$st row to the mth. For each such bond we set $\Psi(n) = \exp ipn$ and seek a Γ, such that $\sum_{n'} e^{2\beta J\delta(n,n')}\Psi(n') = e^{\Gamma(p)}\Psi(n)$; we straightaway obtain $\Gamma(p) = \log(e^{2\beta J} - 1 + q\delta_{p,0})$. Then, the difference between the larger $\Gamma(0)$ and the smaller $\Gamma(p \neq 0)$ is $\Delta\gamma = \log(\frac{e^{2\beta J}-1+q}{e^{2\beta J}-1})$, ultimately to be compared to the difference between the maximum and minimum exponent in V_2, i.e. to $2\beta J$.

[dd]For $q = 2$, $S = \pm 1$, $2\delta(n, n') = 1/2\{(S+1)(S'+1) + (S-1)(S'-1)\} = SS' + 1$. Thus, at $q = 2$ the Potts interaction is an Ising interaction + irrelevant additive constant.

[ee]F.Y. Wu, *Reviews of Modern Physics* **54**, 235 (1982).

[ff]P. Martin, *Potts Models and Related Problems in Statistical Mechanics*, World Scientific, Singapore 1991.

Table 8.3. Critical Exponents in 2D Potts Models.

Type	α	β	γ	δ	ν	η	kT_c/J
$q = 2$	0	1/8	7/4	15	1	1/4	2.27
$q = 3$	1/3	1/9	13/9	14	5/6	4/15	1.99
$q = 4$	2/3	1/12	7/6	15	2/3	1/4	1.82

Duality comes about by the interchange of V_1 with V_2. If we take $n_j - n_{j+1}$ to be a new variable s_j such that the eigenfunctions of V are expressed in the s's, then p_j is replaced by $i(\frac{\partial}{\partial s_j} - \frac{\partial}{\partial s_{j-1}})$. A final substitution $s_j \Leftrightarrow -i\partial/\partial s_j$ results in the effective interchange of V_1 and V_2, mapping the low-T region onto the high-T region. Then, *if there is* a phase transition the two expressions have to coïncide at a point that we identify as T_c:

$$\log\left(\frac{e^{2\beta_c J} - 1 + q}{e^{2\beta_c J} - 1}\right) = 2\beta_c J, \quad \text{i.e.} \quad kT_c = \frac{2J}{\log(1 + q^{\frac{1}{2}})}. \tag{8.55}$$

It is further possible to prove the phase transition in 2D is second-order for $q = 4$ and first-order for $q > 4$. The critical indices in the second-order transitions for the $q = 3$ and 4 models have been calculated precisely and are as given in Table 8.3. For $q = 2$ the model coïncides with the 2D Ising model (cf. Table 8.2).

Again, the conscientious reader will wish to confront Eqs. (8.51) with these data. Ultimately, the importance of the Potts model rests in its generalizations to large q in 3D. Its solutions, even if approximate, can be used to interpret real alloying phase diagrams[ee] such as in Fig. 3.7, more-or-less from "first principles".

Chapter 9

Some Uses of Quantum Field Theory in Statistical Physics

9.1. Outline of the Chapter

This final chapter spins several apparently disconnected threads into a common skein. They include *diffusion-limited* reactions, the Hohenberg–Mermin–Wagner theorems (proving the lack of long-range order in low-dimensions), thermodynamic Green functions and their equations of motion as popularized in the 1960's and finally, the application of this methodology to the study of *glass* — the ultimate random medium.

In the first instance, what starts as an essentially classical topic is best understood in the terminology of a modified Bose–Einstein quantum field theory (QFT). First formulated by Doi in the 1970's,[a,b] the quantum-field-theoretic formulation of the *master equation* of diffusion-limited processes allows a systematic investigation of any number of nonequilibrium events, ranging from fluorescence to explosions. In the remaining instances some applications of QFT to statistical physics are noted, including a theory of glass from the point of view of a quasi-elastic medium in which the normal modes — phonons — are scattered at random.

We begin by examining *diffusion* — the macroscopic consequence of multiple microscopic random walks. For simplicity we shall assume all processes occur on *hypercubic* lattices.[c]

9.2. Diffusion on a Lattice: Standard Formulation

In all dimensions $d \geq 1$, it is *diffusion* (*alias* Brownian motion) that brings the reactants into contact and nourishes the reactions. Let us first consider

---

[a]M. Doi, *J. Phys.* **A9**, 1465 and 1479 (1976).
[b]References and review in D. C. Mattis and M. L. Glasser, *Rev. Mod. Phys.* **70**, 979 (1998).
[c]Continuum theories can be derived by proceeding to the limit $a \to 0$.

the example of a single entity (particle, vacancy, or whatever) diffusing on the N vertices of a space lattice. For simplicity, we restrict random hopping from any given point R to just its immediate nearest-neighbors situated at $R + \delta$. Given $P(R|t)$ as the probability that R is occupied at time t, the *master equation for diffusion* involves the evolution of probabilities and takes the form:

$$\partial_t P(R|t) = -D \sum_\delta \{P(R|t) - P(R + \delta|t)\}$$

$$= -D \sum_\delta \{1 - e^{\delta \cdot \partial/\partial R}\} P(R|t). \tag{9.1}$$

The *rhs* follows from a Taylor series expansion in δ, as is well-known in condensed matter physics.[d] The special solution to this equation, subject to an initial condition $P(R|0) = \delta(R - R_0)$ at $t = 0$,[e] is commonly denoted the *diffusion Green function* or propagator $G(R - R_0|tD)$. Now suppose $Q(R_0)$ to be the probability of the particle being initially at R_0. Then the general solution of (9.1) is just

$$P(R|t) = \sum_{R_0} G(R - R_0|tD) Q(R_0).$$

The *dispersion relation* is found by solving (9.1) using a plane wave with imaginary frequency, i.e. $\exp(-\omega t + i\mathbf{k} \cdot \mathbf{R})$. It is,

$$\omega(\mathbf{k}) = D \sum_\delta (1 - e^{ik \cdot \delta}). \tag{9.2}$$

On hypercubic lattices in d dimensions the vertices (lattice points) are at $\mathbf{r} = a(n_1, n_2, \ldots, n_d)$ where the n's are integers. In this simple case the dispersion $\omega(\mathbf{k})$ is separable, i.e. $\omega(\mathbf{k}) = \sum_{j=1}^d \omega(k_j)$, with $\omega(k_j) = 4D \sin^2(k_j a/2)$ along the jth axis. At long wavelengths, $\omega(\mathbf{k}) \approx Dk^2 a^2$, where a is the lattice parameter. Because D has dimensions sec^{-1} in our units, $D \times a^2 =$ the usual *continuum* diffusion coefficient, that we denote D'.

To construct the hypercubic lattice Green function at $r = R - R_0$ one averages over all admissible plane waves whose k's are restricted to the first Brillouin Zone (BZ) of the lattice. This reduces G to a Kronecker delta at $t = 0$. In the thermodynamic limit the sum over k is replaced by an integral. Given the definition of the Bessel functions of imaginary argument, $I_n(z)$, one expresses G on the hypercubic lattices, making the dependence on D

---

[d] I.e. given a differentiable function $f(x)$, $f(x + a) \equiv e^{a\partial/\partial x} f(x) \equiv e^{iap_x/\hbar} f(x)$.
[e] The same as requiring, with unit probability, that the particle start out at a specified R_0.

explicit henceforth, as follows:

$$G(R|tD) = \frac{1}{N} \sum_{k \subset BZ} e^{(ik \cdot R - tD\omega(\mathbf{k}))}$$

$$= \prod_{j=1}^{d} \frac{a}{2\pi} \oint dk_j \cos k_j n_j e^{-tD\omega(k_j)}$$

$$= e^{-2tDd} \prod_{j=1}^{d} I_{n_j}(2tD) \tag{9.3}$$

At large R and t, G given in Eq. (9.3) smoothly reduces to the continuum $G \equiv (4\pi tD')^{-d/2} \exp{-(R^2/4tD'a^2)}$ so that the probability of finding a particle within a radius $R \propto \sqrt{t}$ from the origin is $O(t^{-d/2})$, from which it might be inferred that, ultimately, particle density decays as $t^{-d/2}$ owing to various causes such as recombination, etc. But this is not always the case.

Consider the interesting case of particle-antiparticle recombination (electron-hole recombination in an amorphous semiconductor where each species performs a random walk to find the other, or vacancy-interstitial recombination), when initially there are equal numbers of both species the density of each decays as t^{-1} in high dimensions ($d \geq 4$) and as $t^{-d/4}$ in $d \leq 4$.[f]

Note that the one-body diffusion Green function is directly generalizable to *many-body* diffusion. For example, in the case of 1 particle of each of 2 diffusing species, the two-body Green function is given as the *product* of the two respective one-body Green functions, by

$$G(\mathbf{r}, \mathbf{r}'|tD_1, tD_2) = \frac{1}{N^2} \sum_{\mathbf{k}, \mathbf{k}' \subset BZ} e^{(ik \cdot r + ik' \cdot r' - t(D_1\omega(\mathbf{k}) + D_2\omega(\mathbf{k}')))} .$$

(This expression resembles a *one*-body Green function on a 2d-dimensional lattice). The Green functions for $n > 2$ particles can be obtained from this, by induction, as higher products of n one-body Green functions.

Diffusion Green functions are normalized, i.e. $\sum_R G(R|tD) = 1$, at $t \geq 0$. They also satisfy a *convolution theorem*. Using the Kronecker lattice delta-function,

$$\frac{1}{N} \sum_{\mathbf{k} \subset BZ} e^{i\mathbf{k} \cdot \mathbf{R}} = \begin{cases} 1 & \text{for } \mathbf{R} = 0 \\ 0 & \text{for } \mathbf{R} \neq 0 \end{cases}$$

[f]D. Toussaint and F. Wilczek, *J. Chem. Phys.* **78**, 2642 (1983).

one derives,

$$\sum_{R'} G(R_0 - R'|t_1 D_\alpha) G(R' - R_2|t_2 D_\beta) = G(R_0 - R_2|t_1 D_\alpha + t_2 D_\beta). \quad (9.4)$$

9.3. Diffusion as Expressed in QFT

The field-theoretic expression for the dimensionless operator part of the *rhs* of (9.1) is the Hermitean operator Ω,

$$\Omega = \sum_R \sum_\delta \{a_R^* a_R - a_{R+\delta}^* a_R\} = \sum_R \sum_\delta (a_R^* - a_{R+\delta}^*) a_R. \quad (9.5)$$

It graphically expresses the "hopping"[g] of a particle from R to $R + \delta$ (and operates on a probability operator $|\Psi(t)\rangle$ that now replaces $\mathcal{P}$. The following equation (not unlike the Schrödinger equation), replaces (9.1):

$$\partial_t |\Psi(t)\rangle = -D\Omega|\Psi(t)\rangle. \quad (9.6)$$

Along with this equation comes an arcane set of rules. Consider the boson operators: the raising (a_R^*) and lowering operators (a_R) at every site R. The usual commutation relation $[a_R, a_{R'}^*] = \delta_{R,R'}$ holds.[h] With $|0\rangle$ defined as the "particle vacuum" annnihilated by a (i.e. $a|0\rangle \equiv 0$), we define the basis states $|n\rangle$ as follows:

$$|n\rangle \equiv (a^*)^n |0\rangle, \quad \text{i.e.} \quad a|n\rangle = n|n - 1\rangle. \quad (9.7)$$

Each $|n\rangle$ is an eigenstate of $a^* a$ with eigenvalue n. Under a simple linear similarity transformation parametrized by an arbitrary constant A, an arbitrary function F of the operators a and a^* transforms as follows:

$$e^{Aa} F(a^*) e^{-Aa} = F(a^* + A) \quad \text{and} \quad e^{-Aa^*} F(a) e^{-Aa^*} = F(a - A). \quad (9.8)$$

If the similarity transformation has a diagonal quadratic exponent, an arbitrary function F of a and a^* transforms as:

$$e^{Aa^* a} F(a, a^*) e^{-Aa^* a} = F(ae^{-A}, a^* e^A). \quad (9.9)$$

This last is a *rescaling* transformation: $a \to Ca$ and $a^* \to a^*/C$.

The careful reader will observe that although a and a^* were defined as boson operators, they *not* Hermitean conjugates of each other and the left

---

[g]Diagonal terms in $a_R^* a_R$ correspond to 1 and $a_{R+\delta}^* a_R$ to $e^{\delta \partial / \partial x}$ in (9.1). Thus, the eigenalues of $D\Omega$ coincide with the previously calculated $\omega(\mathbf{k})$'s.
[h]Hence for most purposes a can be represented by $\partial / \partial a^*$.

eigenstates of Ω are not simply related to its right eigenstates. The norma-
lization of states $|n\rangle$ cannot be quite the same as in the quantum mechanics
of bosons. Given that the operators are meant to deal with essentially
classical "particles" (e.g. interstitial atoms) that propagate on the given
space lattices, it should not be surprising that there are discrepancies.

The essence of what we wish to accomplish is summarized in the *definition*
of the right-hand state $|\Psi(t)\rangle$ as *generating function* for the probabilities
$P(n|t)$. At any given site, it is:

$$|\Psi(t)\rangle \equiv \sum_{n=0}^{\infty} P(n|t)|n\rangle = \sum_{n=0}^{\infty} P(n|t)(a^*)^n|0\rangle \qquad (9.10)$$

$P(n|t)$ is the probability of n particles on this site. The projection operators
with which to recover the P's are $< 0|a^n/n!$ Thus,

$$\left\langle 0 \left| \frac{a^n}{n!} \right| \Psi(t) \right\rangle \equiv \sum_{m=0}^{\infty} P(m|t) \left\langle 0 \left| \frac{a^n}{n!} \right| m \right\rangle = P(n|t). \qquad (9.11)$$

To test the norms we introduce a *reference* state for this site:

$$\langle \lambda | \equiv \langle 0 | e^a \quad \text{(having the property that } \langle \lambda | n \rangle = 1, \text{ all } n.) \qquad (9.12)$$

Then for *any rh* state of the type defined in (9.10) the *normalization* is, and
remains, independent of time:

$$\langle \lambda | \Psi(t) \rangle = \sum_n P(n|t) = 1. \qquad (9.13)$$

The reference state $\langle \lambda |$ is itself a special case of the *Glauber state*
$\langle 0 | e^{\alpha a}$, a left eigenstate of the operator a^* with eigenvalue α. Thus, given
any normal-ordered function F of a and a^*,

$$\langle \lambda | F(a^*, a) = \langle \lambda | F(1, a) = \langle 0 | F(1, a) e^a. \qquad (9.14)$$

For example, given two arbitrary integers $s, q \geq 0$:

$$\langle \lambda | a^{*s} a^q) | \Psi(t) \rangle = \sum_{n>q} \frac{n!}{(n-q)!} P(n|t) \equiv q! \left\langle \binom{n}{q} \right\rangle \qquad (9.15)$$

a function of q and of time t but independent of s.

It is easy to find averages of operator quantities, as follows:

$$\langle \lambda | F(a^* a) | \Psi(t) \rangle \equiv \sum_n F(n) P(n|t) \equiv \langle F \rangle(t). \qquad (9.16)$$

Next, consider X species (individually labeled by $x = 1, 2, \ldots, X$), and
solve for their joint time-dependence. The multispecies generalization of

Eq. (9.6) is $\partial_t |\Psi(t)\rangle = -\sum \mathcal{D}_x \Omega_x |\Psi(t)\rangle$. This first-order *pde* has the explicit solution:

$$|\Psi(t)\rangle = e^{-t \sum_x \mathcal{D}_x \Omega_x} |\Psi(0)\rangle \qquad (9.17)$$

in which the *diffusion operators* Ω_x are:

$$\Omega_x = \sum_R \sum_\delta \{a^*_{x,R} a_{x,R} - a^*_{x,R+\delta} a_{x,R}\} = \sum_k \omega_x(k) c^*_x(k) c_x(k) \qquad (9.18)$$

(with $a^*_R = \frac{1}{\sqrt{N}} \sum_k e^{ik \cdot R} c^*(k)$ and $c^*(k) = \frac{1}{\sqrt{N}} \sum_R e^{-ik \cdot R} a^*_R$.) The c's are the Fourier transforms of the a's. $\omega_x(k)$ is given in (9.2) with D set $= 1$, but is often approximated by its leading term, $\omega(\mathbf{k}) \approx k^2 a^2$.

Next, let us just examine diffusion of a single particle of species x. At $t = 0$, $|\Psi(0)\rangle = a^*_R |0\rangle$ (omitting the subscript x). This reads as: "one particle (of species x) is initially at R".

Use (9.16) to find the time-dependent state, substituting the plane wave operators c for that of the localized particle. Inverting the transformation and invoking Eq. (9.1) to recover D we find:

$$|\Psi(t)\rangle = \sum_{R'} G(R - R'|tD_x) a^*_{x,R'} |0\rangle \qquad (9.19)$$

exactly as expected. (See Problem below.) In this example, the multi-site, 1-species reference state is:

$$\langle \Lambda| \equiv \langle 0| e^{\sum_R a_R} = \langle 0| e^{\sqrt{N} c(0)} . \qquad (9.20)$$

According to (9.16), the average occupancy of any given site R at t is $\langle \Lambda| a^*_{R'} a_{R'} |\Psi(t)\rangle = \langle \Lambda| a_{R'} |\Psi(t)\rangle = G(R - R'|tD)$, again just what might be expected.

Problem 9.1.   (A) Derive (9.19) explicitly, with the aid of the first line in Eq. (9.3). (B) Assuming the initial configuration consisted of one x-particle at each of M sites: $R_1, R_2, R_3, \ldots, R_M$, (again omitting subscript x) show that at $t > 0$,

$$|\Phi(t)\rangle \equiv \sum_{R'_1, R'_2, R'_3, \ldots, R'_M,} \prod_j \{G(R_j - R'_j|tD_x) a^*_{R'_j}\} |0\rangle . \qquad (9.21)$$

(C) Must this result be modified if some of the R_j's coïncide, i.e. if there are initially *two or more* particles at some of the sites?

In some instances it may be simpler to work in the k representation. The multi-species reference state on the N given sites is then,

$$\langle\Lambda| \equiv \langle 0| \prod_{x,R} e^{a_{x,R}} = \langle 0| e^{\sqrt{N}\sum_x c_x(0)} \tag{9.22}$$

and is seen to involve only the $k = 0$ mode for each species.

9.4.   Diffusion *plus* One-Body Recombination Processes

Taking $|\Psi(t)\rangle$ to be the generalization of the generating function (9.10) to numerous sites and species, the master equation becomes:

$$\partial_t |\Psi(t)\rangle = -H|\Psi(t)\rangle \tag{9.23}$$

in which H includes *all* reaction processes *plus* diffusion.

Thus, for pure one-body decay at each site, one would include in H terms such as

$$H_\tau \equiv \sum_x \frac{1}{\tau_x} \sum_R (a_{x,R}^* a_{x,R} - a_{x,R})\,.$$

This H_τ commutes with the diffusion operator because at $k = 0$, $\omega_x(0) = 0$ in the latter. Thus the total wave operator simply factors as,

$$|\Psi(t)\rangle = e^{-tH}|\Psi(0)\rangle = e^{-tH_\tau}(e^{-t\sum_{x,k} H_x(k)}|\Psi(0)\rangle)\,, \tag{9.24}$$

where

$$H_x(k) = D_x \Omega_x = D_x \sum_k \omega_x c_x^*(k) c_x(k)$$

as before. We define $e^{-t\sum_{x,k} H_x(k)}|\Psi(0)\rangle = |\Phi(t)\rangle$, previously calculated in Eq. (9.21).

Φ is a state in which the members of each species have been rearranged as a function of t by the process of diffusion while their respective total numbers remain constant.

To compute the decay in the number of the x'th species due to the τ process we evaluate $\langle n_{x,R}\rangle$, i.e.

$$\left\langle \frac{1}{N}\sum_R n_{x,R} \right\rangle = \langle\Lambda| \frac{1}{N}\sum_R n_{x,R}|\Psi(t)\rangle = \frac{1}{\sqrt{N}}\langle\Lambda|c_x(0)|\Psi(t)\rangle$$

$$= \frac{1}{\sqrt{N}}\langle 0| e^{\sqrt{N}c_x(0)} c_x(0)|\Psi(t)\rangle\,.$$

So,

$$\langle n_{x,R}\rangle(t) = \frac{1}{\sqrt{N}}\langle 0|e^{\sqrt{N}c_x(0)}c_x(0)e^{-\frac{t}{\tau_x}[c_x^*(0)c_x(0)-\sqrt{N}c_x(0)]}|\Phi(t)\rangle\,.$$

After the diffusion operator is commuted all the way to the left, it drops out and the expression simplifies to:

$$\langle n_{x,R}\rangle(t) = \frac{e^{-t/\tau_x}}{\sqrt{N}}\langle 0|c_x(0)|\Psi(0)\rangle = e^{-t/\tau_x}\langle n_{x,R}\rangle(0)\,. \qquad (9.25)$$

This simple result — exponential decay — should not be surprising. $|\Psi(t)\rangle$ is merely the generating function for the probabilities in a model of radioactive decay that we examined earlier, in Sec. 7.6.[i] Although the model is here generalized to the case of diffusing particles, the total decay is independent of where the particles are situated and therefore *diffusion is irrelevant.*

Problem 9.2.   Calculate the variance in n and check whether it too agrees with the calculation in Sec. 7.6 for particles decaying *in situ.*

9.5.   Diffusion and Two-Body Recombination Processes

Next, we allow x to span 2 species A and B that recombine at a rate $1/\tau$ whenever they occupy the same site. The resulting particle C (if any) is then immediately swept out of the system. In the case of A, B annihilation no particle ensues but assume that any energy released is also immediately swept out of the system. The zero-range recombination operator is

$$H_\tau = \frac{1}{\tau}\sum_R (a_R^* a_R b_R^* b_R - a_R b_R)\,.$$

Upon including the diffusion operator, the time development of $|\Psi(t)\rangle$ is given by the master equation $\partial_t|\Psi(t)\rangle = -H|\Psi(t)\rangle$, with $H = H_D + H_\tau$. Here the c_A's are Fourier transforms of the a's, etc.

$$H = \sum_k \omega(k)[D_A c_A^*(k)c_A(k) + D_B c_B^*(k)c_B(k)]$$

$$+ \frac{1}{\tau}\sum_R (a_R^* a_R b_R^* b_R - a_R b_R)\,. \qquad (9.26)$$

---

[i]And also solved by operator methods such as used here, on pp. 981–983 paper cited in footnote b.

A minimal solution of this model requires finding the decay of each species and calculating their correlation functions. Now, if only average quantities $(k = 0)$ were involved in the final results we could just solve the simple mean-field equation. For equal $n_A = n_B$ the master equation reduces to $\partial_t n_A = -(1/\tau)n_A^2$, having the solution $n_A = \tau/(t + t_0)$, where $n_A(0) = \tau/t_0$ defines t_0 and the asymptotic value of n_A is τ/t at large t. This last is, in fact, the correct asymptotic behavior in dimensions $d \geq 4$.

More generally, the solution of the master equation is,

$$|\Psi(t)\rangle = e^{-tH}|\Psi(0)\rangle, \qquad (9.27a)$$

although a second, equivalent, *time-ordered* solution, is more adaptable to a perturbation-theoretic expansion in powers of $1/\tau$:

$$|\Psi(t)\rangle = e^{-tH_D}T\{e^{-\int_0^t dt' H_\tau(t')}\}|\Psi(0)\rangle. \qquad (9.27b)$$

Time-dependence is defined by $op(t) = (\exp tH_D)op(\exp -tH_D)$. The time-ordering operator T ensures that in the expansion of (9.27b), operators at later times are written to the left of those at earlier times.

For $d < 4$, when the fluctuations dominate, the problem becomes quite difficult. One analyzes the problem qualitatively as follows: in neighborhoods where the densities are not exactly equal due to normal fluctuations, recombination exhausts the minority species rapidly. This leaves a surfeit of majority species; for these to disappear they must diffuse to distant regions where the population imbalance is the reverse.

For this reason one names this process "diffusion limited" recombination. Although it still satisfies a power law, the asymptotic rate of decay has to be far slower than in the mean-field limit and thus the exponent must be quite smaller.

The exponent $d/4$ in the asymptotic behavior $\langle n \rangle \propto 1/t^{d/4}$ for 2-body 2-species recombination in $d < 4$ was first found numerically and semi-phenomenologically in Ref. f. The reader is referred to this paper for a more complete explanation. The field-theoretical solution of the model is broached in Ref. b, where numerous references to more recent literature can be found.

Problem 9.3.   *One*-species *two*-body recombination is modeled by a recombination operator

$$H_\tau = \frac{1}{2\tau} \sum_R (a_R^{*2} a_R^2 - a_R^2),$$

similar to the preceding two-species operator. However, in $d < 4$ the density decays much faster asymptotically in the diffusion-limited one-species model than in the two-species model discussed in the text. After musing on this point and referring back to Sec. 9.2, argue persuasively why an asymptotic behavior $1/t^{\gamma(d)}$ with $\gamma(d) > d/4$ is appropriate in the one-species case. *What is* the value of γ in 1D?

9.6.   Questions Concerning Long-Range Order

In the concluding sections of Chapter 4 we analyzed the Bragg (X-Ray) spectrum of lattices at finite T and remarked on the Debye–Waller exponent that diverged in $d \le 2$ dimensions. This indicated the lack of long-range order (LRO) in low dimensional systems that suffer an infrared divergence in the numbers of phonons.

The absence of LRO is easiest to understand in 1D. Consider a chain of atoms located at discrete points x_n, where $x_{n+1} = x_n + a + \Delta_n$ and Δ_n is a random variable caused either by thermal fluctuations or some mismatch of bonds. If the root cause is thermal, Δ will be distributed according to a probability $P(\Delta)$, such that $\langle \Delta \rangle = 0$ and $\langle \Delta^2 \rangle = \alpha^2 T$, with α a small constant. If due to a mismatch, we shall assume $\Delta_n = \pm \Delta_0$, where the $\pm$ are distributed at random and Δ_0 is an arbitrary parameter. Either way, we find for the distance R_m and its variance, between atoms at n and $n + m$,

$$R_m = ma + \sum_{j=n}^{n+m-1} \Delta_j \quad \text{and} \quad \langle (R_m - \bar{R}_m)^2 \rangle = m \langle \Delta^2 \rangle. \tag{9.28}$$

The latter grows with m regardless of the "smallness" of the fluctuations. The effective correlation distance ξ is found by asking: at what m do the *rms* fluctuations exceed the nominal lattice parameter a, making it uncertain whether the last atom is the $n + m$th or the $n + m \pm 1$st This suggests that correlations decay exponentially as $\exp -|R_m|/\xi$ with $\xi = a^3/\langle \Delta^2 \rangle$.

In two dimensions, a logarithmically divergent Debye–Waller exponent does suggest the lack of LRO in *elastic* two-dimensional solids at finite T although the discrete 2D Ising model does have LRO at all finite $T < T_c$.

The issue of LRO in 2D is not clear-cut, as some models have it in 2D and some do not. Is there a relation between thermal fluctuations in quantum systems and LRO?

Building on inequalities developed by Bogolubov and also by Hohenberg, Mermin and Wagner[j] finally resolved this conundrum by proving that — if there is a continuous local symmetry — there can be no LRO at any finite T in dimensions $d \leq 2$.

Their theorem applies not just phonons but also to the "X-Y-Z" Heisenberg models of ferromagnetism ($J < 0$) and anti-ferromagnetism ($J > 0$). But, given that in both models $T_c = 0$ in 1D and 2D, they are obviously in their high-T disordered phases at any finite T and therefore the Mermin–Wagner theorem is redundant!

Still, this theorem does have a nontrivial consequence in the "X-Y" model[k] in 2D. The X-Y model exhibits a continuous phase transition (known as the Kosterlitz–Thouless transition[l]) at a *finite* critical temperature T_c. In this instance the theorem proves[m] (and independent calculations have confirmed), that not only is the high-temperature phase disordered, but so is the low-temperature phase — similar to vapor-liquid phase transitions in 3D.

But the 2D Ising model and the 2D q-state Potts models (and the corresponding lattice gases) behave quite differently. Both have a phase transition at finite T_c (at small q) and both *do* exhibit LRO in the low-temperature phases.

How do the differences in the low-temperature behavior arise? The Ising and Potts models differ from the X-Y and Heisenberg models only in that they have *discrete* local symmetries; their excited states are separated from the ground state by a finite energy gap Δ_g. It is reasonable then to expect that the ground state LRO persists throughout the low temperature regime, $kT < O(\Delta_g)$. In the following we prove the Mermin–Wagner theorem and examine its consequences for systems with continuous local symmetry.

[j]N. D. Mermin and H. Wagner, *Phys. Rev. Lett.* **17**, 1133 and 1307 (1966).

[k]Also known as the "plane-rotator" as the interactions among neighbors involve only the components S_x and S_y in the x-y plane.

[l]J. Kosterlitz and D. Thouless, *J. Phys* **C6**, 1181 (1973). See also: V. L. Berezinskii, *Sov. Phys.* (*JETP*) **32**, 493 (1971) and D. Mattis, *Phys. Lett.* **104**, 357 (1984).

[m]See Problem 9.5 below.

9.7. Mermin–Wagner Theorem

This theorem, stating the absence of spontaneous LRO for systems with a continuous local symmetry in 1D and 2D, takes the form of an inequality. In the presence of an external field (homogeneous field B for a ferromagnet or a *staggered* magnetic field B_{st} in an antiferromagnet), one can show that the relevant order parameter[n] σ is bounded as follows:

$$0 \le \sigma < \left(\frac{T_1}{T}\right)^{2/3} \left(\frac{B}{B_1}\right)^{1/3} \quad \text{in 1D}, \tag{9.29}$$

where T_1 and B_1 are constants, and

$$0 \le \sigma < \left(\frac{T_2}{T}\right)^{1/2} \left|\log\left|\frac{B_2}{B}\right|\right|^{-1/2} \quad \text{in 2D}, \tag{9.30}$$

where T_2 and B_2 are different constants.

No matter how low the temperature, these formulas predict σ vanishes in zero-field.[o] Hence there is no *spontaneous* symmetry-breaking and no LRO.

The proof comes in two stages. For now, we make use of *Bogolubov's inequality*,

$$\left(\frac{1}{2}\langle\{A, A^+\}\rangle_{TA}\langle[[C, H], C^+]\rangle_{TA}\right) \ge (kT|\langle[C, A]\rangle_{TA}|^2) \tag{9.31}$$

to prove (9.29) and (9.30); we derive this inequality in the next section. Here $\{,\}$ is an anticommutator and $[,]$ a commutator.

Isotropic Heisenberg Hamiltonian. Let us illustrate using this simple case. It is governed by the Hamiltonian,

$$H = \sum_{i,j} J(R_{ij})S_i \cdot S_j - B\sum_i S_{j,z}e^{iK \cdot R_j} \tag{9.32}$$

where each quantum spin obeys the usual relations for angular momentum: $[S^-, S^+] = 2\hbar S_z, [S_z, S^\pm] = \pm\hbar S^\pm$ and $S^2 = \hbar^2 s(s+1)$ with $s = 1/2, 1, \ldots$ Spin operators on different sites commute. Next, pick

$$C \equiv S^+(k) = \sum_j e^{-ik \cdot R_j}S_j^+ \quad \text{and} \quad A = S^-(-k - K). \tag{9.33}$$

---

[n] $\sigma \propto$ the magnetization in a ferromagnet or, in a two-sublattice antiferromagnet, the difference of the two antiparallel sublattice magnetizations.

[o] Note that if the *inequalities* were replaced by *equalities*, it would imply $\chi \to \infty$ in zero-field at all T (recall $\chi \propto \partial\sigma/\partial B$). At high T this is clearly impossible! It follows that the "<" signs in (9.29) and (9.30) have to be replaced by "≪" at sufficiently high T.

The three important quantities are: the order parameter,

$$\sigma = \frac{1}{N} \sum_j e^{iK \cdot R_j} \langle S_{j,z} \rangle_{TA} , \qquad (9.34)$$

and the anticommutator,

$$\frac{1}{2} \langle \{A, A^+\} \rangle_{TA} = \frac{1}{2} \langle \{S^+(k+K), S^-(-k-K)\} \rangle_{TA}$$

which is to be summed over k and bounded from above:

$$\frac{1}{N} \sum_k \frac{1}{2} \langle \{S^+(k,K), S^-(-k-K)\} \rangle_{TA}$$

$$= \frac{1}{N} \sum_j (S_{x,j}^2 + S_{y,j}^2) < s(s+1). \qquad (9.35a)$$

Thus,

$$s(s+1) > \frac{kT}{N} \sum_k \frac{|\langle [C, A] \rangle_{TA}|^2}{\langle [[C,H], C^+] \rangle_{TA}} . \qquad (9.35b)$$

The numerator on the *rhs* of (9.35b) is precisely,

$$|\langle [C, A] \rangle_{TA}|^2 = |\langle [S^+(k), S^-(-k-K)] \rangle_{TA}|^2 = (\hbar N \sigma)^2 . \qquad (9.35c)$$

The denominator of (9.35b) is the only quantity involving H and J. It will be replaced by a generous upper bound:

$$\langle [[C,H], C^+] \rangle_{TA} < \frac{N^2}{2} \left(\left(\sum_j |J(R_j)| R_j^2 \right) k^2 + |B\sigma| \right) . \qquad (9.35d)$$

Combining the above, one obtains

$$\sigma^2 \leq \frac{s(s+1)}{2kT} \left(\frac{1}{(2\pi)^d} \int_{BZ} d^d k \frac{1}{(s(s+1) \sum_j |J(R_j)| R_j^2) k^2 + |B\sigma|} \right)^{-1} . \qquad (9.36)$$

The integral diverges for $d \leq 2$ if, in the denominator, one takes the limit $B \to 0$. Then, according to (9.36), $\sigma^2 \equiv 0$, *QED*.

If, instead, one assumes B to be small (but non-zero), it is possible to manipulate the transcendental inequality (9.36) to obtain the Mermin–Wagner inequalities (9.29) and (9.30).

Problem 9.4. Do just that: obtain inequalities (9.29) and (9.30) (to leading orders in B and σ) starting from Eq. (9.36) above with $B \neq 0$.

Problem 9.5. Analyze the X-Y model at low T in the same way as for the preceding Heisenberg model. Because of anisotropy you must now consider two cases: (a) external field B and order parameter in the x-direction or (b) both along the (noninteracting) z-direction. Show there is no LRO at finite T when $B \to 0$.

The Heisenberg and X-Y models govern purely magnetic degrees of freedom and may not be appropriate in metals in which the spin polarized electrons are *itinerant*. In that case, can 1D or 2D *metals* be either ferromagnetic, antiferromagnetic, or sustain spin-density waves or any form of LRO? The answer again is, *no*.

Low-Dimensional Metals. Consider a Fermi sea of interacting electrons in the conduction band of a metal. In terms of the complete set of anticommuting Bloch operators $c_{k,m}$ and $c^+_{k,m}$, the order parameter analogous to that in (9.34) is

$$\sigma = \frac{1}{2}\sum \langle (c^+_{k+K\uparrow}c_{k\uparrow} - c^+_{k+K\downarrow}c_{k\downarrow})\rangle_{TA} . \tag{9.37}$$ [p]

The analogous operators can be chosen as,

$$C \equiv S^+(k) = \sum c^+_{k'\uparrow}c_{k'+k\downarrow} \quad \text{and} \quad A \equiv \sum_{k'} c^+_{k'+k\downarrow}c_{k'-K\uparrow} . \tag{9.38}$$ [q]

The Hamiltonian H consists of three parts: the motional ("kinetic") energy H_0 plus the two-body charge interactions gH' and the magnetic interactions

$$-\frac{B}{2}\sum_k (c^+_{k+K,\uparrow}c_{k,\uparrow} - c^+_{k+K,\downarrow}c_{k,\downarrow}) .$$

The first two contributions to H are modeled by:

$$H_0 = \sum_k \sum_{m=\uparrow,\downarrow} \frac{\hbar^2 k^2}{2m^*} c^+_{k,m}c_{k,m} \quad \text{and} \quad gH' = \frac{1}{2}\sum_{i,j} V(R_{ij})n_i n_j \tag{9.39}$$

---

[p]Use "+" sign for *charge-density* type order parameter.
[q]For *charge-density* studies use a modified C and a similarly modified A, i.e. $C \equiv \rho(k) = \sum_{k',m=\uparrow,\downarrow} c^+_{k',m}c_{k'+k,m}$.

where m^* is the "effective mass" parameter (as obtained from band-structure calculations or experiment) and

$$n_j = \frac{1}{N} \sum_m \sum_{k,k'} e^{i(k'-k)\cdot R_j} c^+_{k,m} c_{k',m}$$

is an occupation-number operator.

Because gH' commutes with the operators A and C in (9.38),[r] for the purposes of inequalities (9.35) we can write

$$H = H_0 - \frac{B}{2} \sum_k (c^+_{k+K,\uparrow} c_{k,\uparrow} - c^+_{k+K,\downarrow} c_{k,\downarrow}).$$

This shows we are dealing with what is effectively a "free electron" system, one that notoriously has *no* tendency for spontaneous symmetry-breaking![s]

Problem 9.6.   With the aid of the σ, A and C appropriate to free electrons, use (9.31) to construct the tightest possible bounds on spontaneous LRO in the free-electron model as defined above in 1D and 2D.

We turn next to the derivation of the crucial inequality (9.31).

9.8.   Proof of Bogolubov Inequality

Define a new type of "inner-product" bracket, (A, D), as follows:

$$(A, D) \equiv \frac{1}{Z} \sum_i \sum_{j \neq i} (i|A|j)^* (i|D|j) \frac{e^{-\beta E_i} - e^{-\beta E_j}}{E_j - E_i}. \qquad (9.40)$$

For an upper bound on the last factor use: $0 < (e^{-ax} - e^{-ay}) \div (y - x) < (1/2)ae^{-a\xi}$, where ξ is x or y, whichever is smaller. It follows that

---

[r]Note to the reader: prove this assertion.
[s]Strictly speaking, this only shows that static *spin-density waves* are excluded in 1D and 2D. However, *charge-density* waves are also completely ruled out in $d \leq 2$ by virtue of similar inequalities applied to slightly modified operators (as indicated in the preceding footnotes).

$(A, A) < \frac{\beta}{2Z} K$, where

$$
K = \sum_i \left\{ \sum_{\substack{j \neq i \\ E_j < E_i}} \langle i | A^+ | j \rangle \langle j | A | i \rangle e^{-\beta E_j} + \sum_{\substack{j \neq i \\ E_j > E_i}} \langle i | A^+ | j \rangle \langle j | A | i \rangle e^{-\beta E_i} \right\}.
$$

A superior bound to K is obtained by removing the restrictions $E_j <$ (or $>$) E_i. Interchange of i, j in the second sum yields,

$$
(A, A) < \frac{\beta}{2} (\langle \{ A, A^+ \} \rangle_{TA}). \tag{9.41}
$$

We invoke a generalized Schwartz' inequality, relevant to the "inner product" (9.40), *viz.*

$$
(A, A)(D, D) \geq |(A, D)|^2 . \tag{9.42}
$$

If one chooses

$$
D = [C^+, H], \quad \text{then } (A, D) = (A, [C^+, H]) .
$$

When this last is inserted into (9.40) the denominator is canceled and one obtains the expression $\langle [C, A] \rangle_{TA}$ appearing on the *rhs* of (9.31). The same holds for (D, D) and so (9.42) reduces to (9.31).

9.9.   Correlation Functions and the Free Energy

Q: What do Bragg scattering, electrical conductivity, magnetism, specific heat and order-disorder phase transitions have in common?

A: All can be analyzed using some sort of correlation function.

At present let us list some correlation functions that have been much discussed over the years. One distinguishes static (e.g. LRO) and time-dependent (*alias* frequency-dependent conductivity $\sigma(\omega)$) correlations. But all can be studied with the aid of some "two-time Green function", as introduced in the following section.

The ultimate goal is an exact or accurate computation of such quantities as, $\langle A(t) B(t') \rangle_{TA}$, where A and B are arbitrary operators, not necessarily distinct. Some tangible examples follow.

In the study of the interacting Bose–Einstein fluid undertaken in Chapter 5, it is generally useful to know correlation functions such as

$$
C_{k,k';q}(g, T) = \langle a^+_{k+q} a^+_{k'-q} a_{k'} a_k \rangle_{TA}
$$

both above and below T_c, given that the thermal average is with respect to an ensemble determined by the full Hamiltonian including H_0 (the kinetic energy part of free bosons) plus the interaction term gH':

$$H = H_0 + \frac{g}{vol} \sum_q v_q \sum_{k,k'} a^+_{k+q} a^+_{k'-q} a_{k'} a_k , \qquad (9.43)$$

with v_q the spatial Fourier transform of the two-body interaction $V(|r_i - r_j|)$. The a's are boson operators subject to the familiar commutation relations, $[a_k, a^+_{k'}] = \delta_{k,k'}$ (all other commutators $= 0$). The Free energy (and from it, a number of other thermodynamic functions) is given *exactly* by,

$$F = F_0 + \frac{1}{vol} \sum_q v_q \int_0^g dg' \sum_{k,k'} C_{k,k';q}(g', T) \qquad (9.44)$$

where F_0 is the (known) free energy of the ideal boson gas of N particles at the given temperature.

How to choose A and B in this $t = t' = 0$ example? There are several strategies that can be adopted: $A = \rho(q) = \sum_k a^+_{k+q} a_k$ and $B = \rho(-q)$ or $A = \sum_{k,k'} a_{-k'} a_k$ and $B = A^+$ are distinct bilinear choices. The results for weakly-interacting bosons in Sec. 5.12 can be generalized after further factorization into corelation functions quadratic in the field operators. This requires factoring $a^+_{k+q} a^+_{k'-q} a_{k'} a_k$ into all possible bilinear factors such as

$$2a^+_{k+q} \langle a^+_{k'-q} a_{k'} \rangle_{TA} a_k + 2a^+_{k+q} a_{k'} \langle a^+_{k'-q} a_k \rangle_{TA}$$

$$+ \langle a^+_{k+q} a^+_{k'-q} \rangle_{TA} a_{k'} a_k + a^+_{k+q} a^+_{k'-q} \langle a_{k'} a_k \rangle_{TA} .$$

If to this approximation we add the assumption that thermal averaging (TA) conserves translational invariance, quantities such as $\langle a^+_{k'-q} a_{k'} \rangle_{TA} = \langle a^+_{k'} a_{k'} \rangle_{TA} \delta_{q,0}$ simplify greatly, as does $\langle a^+_{k'-q} a_k \rangle_{TA} = \langle a^+_k a_k \rangle_{TA} \delta_{k'-q,k}$, and as do the anomalous averages $\langle a_{k'} a_k \rangle_{TA} = \langle a_{-k} a_k \rangle_{TA} \delta_{k',-k}$. All can be calculated self-consistently, after diagonalizing the momentum-conserving quadratic form similar to H_2 in Chapter 5. These self-consistent solutions can be made to yield a mean-field theory of the fluid-superfluid phase transition in $d > 2$. But for $d \leq 2$, owing to the absence of LRO this does not "work" and $\langle a_{-k} a_k \rangle_{TA}$ vanishes.

If one extends this analysis to the 3D *solid phases* of Bose–Einstein particles, there is an even greater number of nonvanishing terms including:

$$\langle a^+_{k'-q} a_{k'} \rangle_{TA} = \sum_n \langle a^+_{k'-Q_n} a_{k'} \rangle_{TA} \delta_{q,Q_n}$$

and the anomalous average

$$\langle a_{k'}a_k\rangle_{TA} = \sum_n \langle a_{-k+Q_n}a_k\rangle_{TA}\delta_{k',-k+Q_n}\,,$$

the sums ranging over vectors Q_n of the solid's reciprocal lattice.

The *electron gas* provides yet another example in which correlation functions play an important rôle. To the Hamiltonian of the free electron gas, expressed in a complete set of anticommuting operators, the c's and c^+'s,

$$H_0 = \sum_{k,m=\uparrow,\downarrow} (\varepsilon(k) - \mu)c^+_{k,m}c_{k,m}\,,$$

we add the Coulomb interaction Hamiltonian,

$$gH' = \frac{4\pi e^2}{vol} \sum_{\substack{q \\ (q_z>0)}} \frac{1}{q^2}\rho(q)\rho(-q)\,, \tag{9.45}$$

in which $\rho = \rho_\uparrow + \rho_\downarrow$ and $\rho_m(q) = \sum_k c^+_{k+q,m}c_{k,m}$. The Coulomb interaction energy scales with[t] $1/r_s$ hence with $k_F a_0$ (a_0 is the physical lattice parameter). The kinetic energy scales with $(k_F a_0)^2$.

Hence, at extremely low density and temperature the Coulomb energy becomes paramount. To minimize it, the translational symmetry has to be broken and the 3D electron gas condenses spontaneously into the antiferromagnetic "Wigner lattice".[u] Because Bragg reflections are part and parcel of any space lattice with LRO, some set of $\rho(Q_n)$ have a nonvanishing expectation value and $\langle \rho(Q_n)\rho(-Q_n)\rangle_{TA} \approx \langle \rho(Q_n)\rangle_{TA}\langle \rho(-Q_n)\rangle_{TA}+$ fluctuations.

At a slightly smaller value of r_s the 3D lattice "melts" into a partly ferromagnetic translationally invariant fluid, such that $\langle S_{z,total}\rangle_{TA} \neq 0$. Thus there is magnetic LRO in the ground state. The order parameter presumably persists over a range of temperature. At the date of writing the dependence of the Curie temperature on the density has not been established.

At smaller r_s still, in the range of densities that are typical of ordinary metals ($k_F a > 1$,) the 3D Coulomb gas behaves very much like a noninteracting, ideal, Fermi–Dirac gas. The relevant correlations are

$$C_q(e^2) \equiv \langle \rho(q)\rho(-q)\rangle_{TA}$$

as in Eq. (9.45) or, more microscopically, $\langle c^+_{k+q,m}c_{k,m}c^+_{k',m'}c_{k'+q,m}\rangle_{TA}$ (which can then be summed to obtain C_q). Using this quantity in conjunction with

---

[t] r_s^3 is $\propto$ volume occupied by each electron $\propto 1/$electron density.
[u] This can be either hexagonal close packed or body-centered cubic, and has lattice parameter $\approx r_s$.

Eq. (9.45) we obtain the exact free energy of the electron fluid as:

$$F = F_0 + \frac{2\pi}{Vol} \sum_q \frac{1}{q^2} \int_0^{e^2} de'^2 C_q(e'^2).$$ (9.46)

In short, correlation functions are ubiquitous. They are closely related to thermodynamic and transport properties and they are valuable tools for the calculation of critical phenomena, critical exponents, etc. The no-LRO theorems in low dimensions were based on bounds on some appropriate correlation functions. Unfortunately, with some exceptions there is no straightforward, self-consistent way to obtain correlation functions *directly* from first principles. We must proceed *indirectly* by solving the equations of motion of the related Green's functions. Next, we examine the connections between the two.

9.10. Introduction to Thermodynamic Green's Functions

Ever since their invention by the 19th Century English mathematician George Green, his eponymous functions have served to solve differential equations and, when this is not possible, to replace them by equivalent but more transparent integral equations that easily incorporate the stated boundary conditions. Students of quantum theory are familiar with this application of Green's functions fundamental to the development of scattering theory. Those who are familiar with aspects of quantum field theory will recognize the usefulness in the many-body perturbation theory of the "time-ordered Green functions", generally known as *propagators*. These are similar to $G(\omega) \equiv \frac{1}{H-\omega}$. Using such G's one can obtain the *density-of-states* $\rho(\omega)$ for any *arbitrary* Hermitean Hamiltonian using the obvious formula: $\rho(\omega) = \frac{i}{2\pi} Tr\{G(\omega - i\varepsilon) - G(\omega + i\varepsilon)\}$.

The *thermodynamic* functions that are the subject of this section are quite special and were developed in the late 1950's specifically to calculate the various correlation functions in many-body problems of interacting particles at a finite T. A number of useful results are summarized in a clear and readable contemporary review by Zubarev.[v]

The Green functions come in 3 genuses or 6 distinct types: *retarded*, *advanced* and *causal*, each of which can be used in conjunction with either

---

[v]D. N. Zubarev, *Sov. Phys. (Uspekhi)* **3**, 320–345 (1960) (English translation of the original in *Usp. Fiz. Nauk* **71**, 71–116 (1960)). Also: G. D. Mahan, *Many-Particle Physics* 2nd Edition, Plenum, New York, 1990, pp. 81–234.

Fermi–Dirac or Bose–Einstein statistics. The following shows the intimate connection between Green functions and their correlation functions:

$$\langle B(t')A(t)\rangle_{TA} = \frac{i}{2\pi} \int_{-\infty}^{+\infty} d\omega e^{-i\omega(t-t')} \frac{G_a^{(\pm)}(\omega) - G_r^{(\pm)}(\omega)}{e^{\beta\omega} \pm 1} . \tag{9.47}$$

The $(\pm)$ superscripts are paired with $(\pm)$ in the denominator. They refer to commutator $(-)$ or anticommutator $(+)$ Green functions, The subscript a refers to "advanced" and r to "retarded", two branches of a single function $G(z)$ in the complex $z = \omega + i\varepsilon$ plane with singularities (poles or a branch cut) exclusively on the real axis. Thus, $G_a(\omega) = G(\omega+i\varepsilon)$ and $G_r(\omega) = G(\omega-i\varepsilon)$. Although simple,[v] the proof of (9.47) is too lengthy to reproduce here.

In the time domain, $G_a^{\pm}(t - t') = i\vartheta(t' - t)\langle[A(t), B(t')]_\pm\rangle_{TA}$ is defined in terms of a commutator $(-)$ or anticommutator $(+)$, $[A(t), B(t')]_\pm = A(t)B(t') \pm B(t')A(t)$. The retarded function is, $G_r^{(\pm)}(t - t') = -i\vartheta(t - t')\langle[A(t), B(t')]_\pm\rangle_{TA}$. Here $\vartheta(t)$ is the Heaviside function (the unit step, 1 for $t > 0$ and zero for $t < 0$). The Fourier transform of both is a common function written more suggestively as $\langle\langle A|B\rangle\rangle_\omega$ or even better, as $G_{A,B}(\omega)$.

The function $G_a^{(\pm)}(t)$ satisfies a simple equation of motion:

$$i\frac{dG_a^{(\pm)}(t - t')}{dt} = \delta(t - t')\langle[A(t), B(t)]_\pm\rangle_{TA}$$

$$- \vartheta(t' - t)\left\langle\left[\frac{dA(t)}{dt}, B(t')\right]_\pm\right\rangle_{TA} \tag{9.48a}$$

where $idA/dt = AH - HA \equiv [A, H]_-$, assuming the time-dependence of operators to be the usual $A(t) \equiv (\exp +itH)A(\exp -itH)$. Here and throughout, we take $\hbar = 1$. Now note the last term in the above equation is also an advanced Green function, of a new type:

$$-\vartheta(t' - t)\left\langle\left[\frac{dA(t)}{dt}, B(t')\right]_\pm\right\rangle_{TA} = i\vartheta(t' - t)\langle[[A(t), H], B(t')]\rangle_{TA} \tag{9.48b}$$

in which $C \equiv AH - HA$ takes the place of A. We could work out *its* equation of motion and iterate indefinitely in this manner, until either there is closure or we perform an arbitrary truncation.

The retarded function satisfies a similar equation. Assuming that equal time commutators are independent of t and that the 2-time commutators are a function of $t - t'$ alone, we can Fourier transform these equations to obtain a single equation of motion for both,

$$\omega\langle\langle A|B\rangle\rangle_\omega = \langle[A, B]_\pm\rangle_\omega + \langle\langle[A, H]|B\rangle\rangle_\omega . \tag{9.49}$$

For operators A, B that *commute* at equal times one generally elects to use the commutator $[A, B] = [A, B]_-$ while for operators A, B that *anticommute* at equal times it is preferable to use $\{A, B\} = [A, B]_+$ instead. The quantity $[A, H]$ on the *rhs* is *always* the commutator $[A, H]_-$.

"A picture is worth a thousand words". To conclude this book we provide the examples below to illustrate some uses of Green functions in many-body physics.

Example 1.   Free Electrons: *time-dependent correlations.*

$$H_0 = \sum_k \sum_m e(k) c_{k,m}^+ c_{k,m}, \quad \text{where } e(k) = \varepsilon(k) - \mu.$$

Let $A = c_{k,m}$ and $B = c_{k',m'}^+$ be fermion operators satisfying the usual anticommutators: $\{c_{k,m}^+, c_{k',m'}\} = \delta_{k,k'} \delta_{m,m'}$. We opt for the *anticommutator* Green functions. Now Eq. (9.49) reads:

$$\omega \langle\langle k, m | k'm' \rangle\rangle_\omega = \delta_{k,k} \delta_{m,m'} + e(k) \langle\langle k, m | k'm' \rangle\rangle_\omega, \tag{9.50}$$

in which only the relevant quantum numbers (k, m) are retained and the c symbols are omitted for brevity. Here closure comes immediately. We solve for Green's function:

$$\langle\langle k, m | k', m' \rangle\rangle_\omega = \frac{\delta_{k,k'} \delta_{m,m'}}{\omega - e(k)}. \tag{9.51}$$

Then, according to (9.47) the correlation function is:

$$\langle c_{k'm'}^+(t') c_{k,m}(t) \rangle_{TA} = \delta_{k,k'} \delta_{m,m'} \frac{i}{2\pi} \int_{-\infty}^{+\infty} d\omega \frac{e^{-i\omega(t-t')}}{e^{\beta\omega} + 1}$$

$$\times \left\{ \frac{1}{\omega + i\gamma - e(k)} - \frac{1}{\omega - i\gamma - e(k)} \right\}. \tag{9.52}$$

In $\lim \cdot \gamma \to 0$ the curly bracket is exactly $-i2\pi\delta(\omega - e(k))$, hence

$$\langle c_{k'm'}^+(t') c_{k,m}(t) \rangle_{TA} = \delta_{k,k'} \delta_{m,m'} \frac{e^{-ie(k)(t-t')}}{e^{\beta e(k)} + 1}.$$

This vanishes unless the quantum numbers are the same. If they are, then we recover the familiar Fermi function at equal times. For $t \neq t'$ the same formula yields the one-particle auto-correlation function.

Problem 9.7.   Verify that the use of a *commutator* Green function with the above choices of fermion operators, A and B would not be quite so simple.

Example 2. Free Bosons: $H_0 = \sum_k e(k) a_k^+ a_k$ (with all $e(k) > 0$, of course.) Let $A = a_k$ and $B = a_{k'}^+$ be *boson* operators satisfying the usual commutation relations. The only nonvanishing commutators are $[a_k, a_{k'}^+] = \delta_{k \cdot k'}$. The equations of motion of the commutator Green functions are:

$$\omega \langle\langle k|k'\rangle\rangle_\omega = \delta_{k,k'} + e(k) \langle\langle k|k'\rangle\rangle_\omega \tag{9.53}$$

having a solution,

$$\langle\langle k|k'\rangle\rangle_\omega = \frac{\delta_{k,k'}}{\omega - e(k)}. \tag{9.54}$$

The correlation function is:

$$\langle c_{k'}^+(t') c_k(t) \rangle_{TA} = \delta_{k,k'} \frac{i}{2\pi} \int_{-\infty}^{+\infty} d\omega \, \frac{e^{-i\omega(t-t')}}{e^{\beta\omega} - 1}$$

$$\times \left\{ \frac{1}{\omega + i\gamma - e(k)} - \frac{1}{\omega - i\gamma - e(k)} \right\}. \tag{9.55}$$

The curly bracket is $-i2\pi\delta(\omega - e(k))$, i.e.

$$\langle c_{k'}^+(t') c_k(t) \rangle_{TA} = \delta_{k,k'} \frac{e^{-ie(k)(t-t')}}{e^{\beta e(k)} - 1} \quad \text{and} \quad \langle n_k \rangle_{TA} = (e^{-\beta e(k)} - 1)^{-1}.$$

Example 3. Scattering of Normal Modes: Let $H = H_0 + gH'$, with H_0 being the Hamiltonian of the free bosons in Example 2; gH' is the scattering from a localized impurity at $R_{\rm I}$, the mass or spring constants of which differ from those of the host in an amount g:

$$gH' = \frac{g}{Vol} \sum_k \sum_{\substack{k' \\ k_z' > k_z}} e^{i(k-k') \cdot R_{\rm I}} \sqrt{e(k) e(k')} (a_{k'} + a_{-k'}^+)(a_k + a_{-k}^+). \tag{9.56}$$

Clearly for the thermodynamic function we need the correlations $\langle (a_{k'} + a_{-k'}^+)(a_k + a_{-k}^+) \rangle_{TA}$ as a function of g; call this $F_{k,k'}(g)$. Choosing the position of the impurity to be the origin, $R_I = 0$, we find:

$$F = F_0 + \int_0^g dg' \frac{1}{Vol} \sum_k \sum_{k'} \sqrt{e(k) e(k')} F_{k,k'}(g'). \tag{9.57}$$

To obtain F we require two distinct Green functions: not just

$$\langle\langle (a_k + a_{-k}^+)|(a_{-k'} + a_{k'}^+) \rangle\rangle_\omega \equiv G_{k,k'}(\omega)$$

but also $\langle\langle (a_k - a_{-k}^+)|(a_{-k'} + a_{k'}^+) \rangle\rangle_\omega \equiv K_{k,k'}(\omega)$, to which G connects.

Then, the coupled equations of motion for G and K are:

$$\omega G_{k,k'}(\omega) = e(k)K_{k,k'}(\omega) \tag{9.58a}$$

and

$$\omega K_{k,k'}(\omega) = 2\delta_{k,k'} + e(k)G_{k,k'}(\omega) + \frac{2g}{Vol}\sum_{k''}\sqrt{e(k)e(k'')}G_{k'',k'}(\omega). \tag{9.58b}$$

Eliminating K in (b) by the use of (a):

$$\omega^2 G_{k,k'}(\omega) = 2e(k)\delta_{k,k'} + e^2(k)G_{k,k'}(\omega) + \frac{2e^{\frac{3}{2}}(k)g}{Vol}\sum_{k''}\sqrt{e(k'')}G_{k'',k'}(\omega).$$

Denote $\frac{1}{Vol}\sum_{k''}\sqrt{e(k'')}G_{k'',k'}(\omega) \equiv \Gamma_{k'}(\omega)$. Then the above has the solution,

$$G_{k,k'}(\omega) = \frac{2e(k)\delta_{k,k'}}{\omega^2 - e^2(k)} + \frac{e^{\frac{3}{2}}(k)}{\omega^2 - e^2(k)}2g\Gamma_{k'}(\omega). \tag{9.59a}$$

This solution is complete once G is inserted into the definition of Γ, whence one obtains:

$$\Gamma_{k'}(\omega) = \frac{2}{Vol} \times \frac{1}{1 - 2gS(\omega)} \times \frac{e^{\frac{3}{2}}(k')}{\omega^2 - e^2(k')}, \tag{9.59b}$$

where $S(\omega) = \frac{1}{Vol}\sum_{k''}\frac{e^2(k'')}{\omega^2 - e^2(k'')}$. To evaluate S as an integral we next introduce the density-of-states function $\rho(e)$ for the eigenstates of H_0. $S(\omega)$ is *inherently* complex; it has a discontinuity (branch-cut) across the real ω axis:

$$S(\omega \pm i\varepsilon) = \mp\frac{\pi i}{2}\omega\rho(\omega) + P \cdot P \cdot \int_0^\infty de\rho(e)\frac{e^2}{\omega^2 - e^2}$$

$$= \mp\frac{\pi i}{2}\omega\rho(\omega) + R(\omega) \tag{9.60}$$

with $P \cdot P$. the abbreviation for a "principal part" integration and $\varepsilon \to 0$; note that $R(0) = -1$. Equation (9.57) yields a free energy

$$F = F_0 + \int_0^g dg' \frac{1}{2Vol}\sum_k\sum_{k'}\sqrt{e(k)e(k')}\frac{1}{2\pi i}\int_{-\infty}^\infty d\omega \frac{1}{e^{\beta\omega} - 1}$$

$$\times \{G_{k,k'}(\omega - i\varepsilon) - G_{k,k'}(\omega + i\varepsilon)\}. \tag{9.61a}$$

After inserting the solutions and combining terms this expression simplifies further:

$$F = F_0 + 2 \int_0^g dg' g' \int_{-\infty}^{\infty} d\omega \frac{1}{e^{\beta\omega} - 1}$$

$$\times \frac{\omega\rho(\omega)}{(1 - 2g'R(\omega))^2 + g'^2(\pi\omega\rho(\omega))^2} \tag{9.61b}$$

Problem 9.8. In order to fully reduce this expression to quadrature, first evaluate the integral over g' in terms of the "phase shift" $\theta(g, \omega) \equiv \arctan \frac{g\pi\omega\rho(\omega)}{1-2gR(\omega)}$; then estimate the leading dependence of $F - F_0 = \Delta F$ on T at low T assuming $\rho \propto \omega^2$.

Example 4. Eigenvalues of Random Matrix. A real, symmetric $N \times N$ matrix is characterized by $1/2N(N + 1)$ independent parameters, an Hermitean matrix by N^2. Assume a symmetric matrix, each of whose matrix elements m_{ij} is real, independently random about a zero mean and has a given, constant, *rms* value $g/\sqrt{N}$. In the thermodynamic limit, what is the distribution of its eigenvalues?

Mathematically, the problem is identical to that of finding the *density-of-states* of an Hamiltonian $H = \frac{g}{\sqrt{N}} \sum_{i,j} \varepsilon_{ij} c_i^+ c_j$ (where $\varepsilon_{ij} = \varepsilon_{ji} = \pm 1$ at random) in the N-dimensional space of occupancy 1. For one particle the commutation relations do not matter and we can choose the c's to be either fermions or bosons. Let us pick *fermions*. Then we write the equation of motion of the appropriate Green function as,

$$\omega \langle\langle d_0 | d_0^+ \rangle\rangle_\omega = 1 + g_1 \langle\langle d_1 | d_0^+ \rangle\rangle_\omega . \tag{9.62a}$$

Here we have taken *any one* of the N c_i's and identified it as d_o. By definition this yields d_1, defined by $[d_0, \frac{1}{\sqrt{N}} \sum_{i,j} \varepsilon_{ij} c_i^+ c_j] \equiv d_1$. Notice that $\{d_1, d_1^+\} = 1 + O(1/N)$. Iterating: the commutator $[d_n, \frac{1}{\sqrt{N}} \varepsilon_{ij} c_i^+ c_j] - d_{n-1} \equiv d_{n+1}$ defines $d_2, d_3, \ldots$. Then, for $n \geq 1$,

$$\omega \langle\langle d_n | d_0^+ \rangle\rangle_\omega = g_n \langle\langle d_{n-1} | d_0^+ \rangle\rangle_\omega + g_{n+1} \langle\langle d_{n+1} | d_0^+ \rangle\rangle_\omega . \tag{9.62b}$$

Each coefficient g_n is the product of g *times* the length of a 1D random walk of N steps each of length $\approx 1/\sqrt{N}$; these, to a first approximation, are all equal to g. (Indeed, the fluctuations disappear in the thermodynamic limit $N \to \infty$.) We also note that Eqs. (9.62) are generic, i.e. independent of the choice of the initial c_i to be identified with d_0 in the thermodynamic limit. We call this type of matrix "self-averaged". Let us now define

$$G(\omega) = Tr\{d_0^+(\omega - H)^{-1}d_0\} \cdot \text{ and its related } dos$$

$$\rho(\omega) \equiv \frac{i}{2\pi}(G(\omega - i\varepsilon) - G(\omega + i\varepsilon)). \tag{9.63}$$

As noted above, $\rho(\omega)$ is independent of the initial choice of d_0. Upon examination of the coefficients in Eqs. (9.62), one identifies the array of coefficients as a tri-diagonal *Toeplitz-type* matrix:

$$\tilde{M} = \begin{pmatrix} 0g00000\dots \\ g0g0000\dots \\ 0g0g000\dots \\ 00g0g00\dots \end{pmatrix},$$

whose eigenvectors are (by inspection) $v(k) = \sqrt{\frac{2}{N}}(\sin k, \sin 2k, \sin 3k, \dots)$, each with a corresponding eigenvalue $\omega = 2g \cos k$. Here the k's have to be chosen to satisfy a boundary condition $\sin kN = 0$. Then, $k = p\pi/N$, where $p = 1, 2, \dots, N$. Evaluating (9.63) in this diagonal representation, we obtain:

$$\rho(\omega) = \frac{2}{N}\sum_k \sin^2 k \, \delta(\omega - 2g \cos k) = \frac{2}{\pi}\int_0^\pi dk \sin^2 k \, \delta(\omega - 2g \cos k)$$

$$= \frac{1}{2\pi g^2}\sqrt{(2g)^2 - \omega^2} \text{ for } \omega^2 < (2g)^2 \text{ (zero otherwise)}, \tag{9.64}$$

known as "Wigner's semicircular dos". In light of this exact result the diligent reader might now wish to revisit his/her solution to Problem 1.6.

A similar methodology has been applied to the thermodynamics of *glasses* by the author and his students,[w] under the assumption that the principal effect of random spatial fluctuations is to scatter plane-wave normal modes of an underlying lattice. A linear dependence of the specific heat $c \propto T$ is found in the theory[w] but only for sufficiently large disorder: $g > g_c$, while Debye's law is recovered for $g < g_c$ at low-T. Experiments have consistently revealed a linear low-T specific heat in *all* glasses and amorphous substances, $c \propto T$. This is a most significant departure from Debye's Law (Chapt. 5) that had firmly established $c \propto T^3$ in all crystals. Thus, our calculation indicate that the glass phase $(g > g_c)$ constitutes a thermodynamic phase that is both distinct and distinguishable from that of the disordered crystal $(g < g_c)$.

[w]M. Molina and D. Mattis, *Phys. Lett.* **A159**, 337 (1991), also J. Yáñez, M. Molina and D. Mattis, *ibid.* **A288**, 277 (2001).

Index